Modeling, Reliability and Health Management of Lithium-Ion Batteries—2nd Edition

Guest Editors
Fei Feng
School of Automation
Chongqing University
Chongqing
China

Rui Ling
School of Automation
Chongqing University
Chongqing
China

Yi Xie
College of Mechanical and Vehicular Engineering
Chongqing University
Chongqing
China

Shunli Wang
Power College
Inner Mongolia University of Technology
Hohhot
China

Jinhao Meng
School of Electrical Engineering
Xi'an Jiaotong University
Xi'an
China

Jiale Xie
Automation Department
North China Electric Power University
Baoding
China

Editorial Office
MDPI AG
Grosspeteranlage 5
4052 Basel, Switzerland

This is a reprint of the Special Issue, published open access by the journal *Batteries* (ISSN 2313-0105), freely accessible at: https://www.mdpi.com/journal/batteries/special_issues/OJVX53396A.

For citation purposes, cite each article independently as indicated on the article page online and as indicated below:

Lastname, A.A.; Lastname, B.B. Article Title. *Journal Name* **Year**, *Volume Number*, Page Range.

ISBN 978-3-7258-6199-6 (Hbk)
ISBN 978-3-7258-6200-9 (PDF)
https://doi.org/10.3390/books978-3-7258-6200-9

Contents

batteries

MDPI

Editorial

Modeling, Reliability, and Health Management of Lithium-Ion Batteries (2nd Edition)—A Summary of Contributions and Future Outlook

Fei Feng

Key Laboratory of Complex System Safety and Control, Ministry of Education, Chongqing University, Chongqing 400044, China; feifeng@cqu.edu.cn

Lithium-ion batteries (LIBs) are a cornerstone technology driving transportation electrification and renewable energy storage systems. However, unlocking their full potential necessitates precisely addressing the challenges associated with their modeling, reliability assurance, and full-lifecycle health management. Following the success of our first edition, we are delighted to present the Second Edition of the Special Issue "Modeling, Reliability, and Health Management of Lithium-Ion Batteries". This Special Issue aimed to gather the latest research advancements in the field. We were honored to receive numerous high-quality submissions, from which we have selected eight outstanding research articles (contributions 1–8).

These papers collectively depict the current frontier of LIB research, and their contributions can be summarized into three interconnected core areas:

1. Advancements in Battery Modeling Depth and Efficiency

Accurate battery models are the foundation of all diagnostic and management strategies. The articles in this Special Issue contribute to enhancing both the efficiency and depth of these models. Savu et al. (contribution 1) proposed novel parameter extraction methods for widely used equivalent circuit models (ECMs) based on linear regression. Their research demonstrates that this approach can reduce computational costs to as low as 1.32% of traditional optimization algorithms while maintaining high accuracy, paving the way for online model implementation in Battery Management Systems (BMSs). The accuracy of ECMs is critical for system-level applications; for instance, Chen et al. [1] utilized ECMs to evaluate the systemic impact of battery cell imbalance on the range of electric vehicles (EVs). Chen et al. (contribution 2) focused on Lithium Titanate Oxide (LTO) batteries, known for their rapid charge/discharge capabilities. They introduced an electrical model based on the solid-phase diffusion equation, which more accurately describes the unique electrochemical characteristics of LTO batteries, especially near the charge/discharge cutoff conditions, achieving a maximum voltage error below 3%. Modeling the aging characteristics of specific materials like LTO is also an active research area; for example, Fang et al. [2] have explored performance simulation methods for LTO batteries based on an aging-effect coupling model.

2. Systemic Protection for Safety, Reliability, and Thermal Runaway (TR)

Ensuring battery safety is one of the most pressing tasks in this field. Several papers in this Special Issue delve into reliability and fault protection from various perspectives. Regarding thermal runaway (TR) mechanisms, Ayayda et al. (contribution 3) developed a coupled model for TR and internal pressure specifically for prismatic cells. They validated the pressure model's effectiveness with experimental data, filling a research gap in the pressure dynamics of prismatic cells during TR. This line of research is critical, as the fusion of internal pressure and temperature sensing is considered key to multi-level TR warnings, a strategy also investigated by Zhu et al. [3]. Lee et al. (contribution 4) linked TR research to the actual operating environments of Energy Storage Systems (ESS). Their evaluation model confirmed that the ambient temperature rise rate and convective heat transfer coefficient are more critical factors influencing TR than the C-rate of charging and discharging. In addition to the operating environment, a review by Chen et al. [4] highlights that internal short circuits triggered by manufacturing defects, particularly metallic foreign matter, represent another fundamental source of TR that must be controlled at the production origin.

On the fault diagnosis front, Fan et al. (contribution 5) proposed a fault diagnosis method for battery packs based on relative entropy and State of Charge (SOC) estimation. This method can promptly detect voltage/temperature sensor faults and internal short circuits, and quantitatively assess the short-circuit resistance, thereby significantly enhancing pack safety. Concurrently, Li et al. [5] explored another effective diagnostic path, employing signal decomposition and two-dimensional feature clustering to achieve robust, early detection of battery faults.

3. Intelligent Prediction of Health Management (SOH) and Aging Mechanisms

Predicting and managing the battery's State of Health (SOH) is crucial for extending its lifespan and enabling second-life applications. This Special Issue showcases the latest progress in data-driven and hybrid physics-data approaches. Yao et al. (contribution 6) introduced a physics-guided machine learning approach that, using data from only the first five cycles, can determine the dominant fading mechanism with 95.6% accuracy and achieve excellent performance in predicting lifetime capacity fade. This concept of fusing physical mechanisms with machine learning is a frontier in SOH prediction, with researchers like Lu et al. [6] also proposing frameworks that fuse coupled degradation mechanisms with machine learning to enhance lifespan prediction accuracy.

Addressing the practical challenge of small-sample data, Liu et al. (contribution 7) developed a Multi-Encoder BHTP Autoencoder for robust SOH prediction. This model demonstrated outstanding performance on the NASA dataset, effectively overcoming interference from battery inconsistency and capacity recovery phenomena. Kalk et al. (contribution 8) focused on practical applications, developing an aging-optimized multi-stage constant current (MCC) fast charging algorithm based on three-electrode measurements. This algorithm successfully reduces the 0–80% SOC charging time by 30% without accelerating aging by avoiding Li-plating, striking an excellent balance between charging speed and battery health. From a more macroscopic, system-level perspective, research by Jia et al. [7] demonstrates that battery aging can also be effectively suppressed by optimizing the "speed trajectory control strategy" of the electric vehicle itself, showcasing the expansion of health management into vehicle control systems.

The eight articles in this Special Issue collectively showcase the latest advancements in LIB modeling, safety, and health management. These findings not only deepen our understanding of the internal mechanisms of batteries but also provide more efficient and robust engineering solutions. Looking ahead, the field is moving towards more integrated and intelligent solutions. We anticipate a convergence of physics-based principles with data-driven tools, such as in Physics-Informed Neural Networks (PINNs), to create high-fidelity digital twins. Indeed, these trends are already emerging. Recent studies have demonstrated digital twin modeling methods driven by data-mechanism fusion and the use of PINNs to accurately predict the degradation trajectories of energy storage systems like supercapacitors [8]. These models will increasingly evolve from electro-thermal coupling to encompass comprehensive "mechano-thermal-electrical" multiphysics. A parallel and crucial trend is the development of data-efficient, lightweight algorithms—leveraging techniques like transfer and federated learning [9]—that can be deployed directly on-BMS for real-time diagnostics. Ultimately, these advancements will support a more holistic, system-level approach to health management, integrating multi-scale data across the entire battery lifecycle, from production to second-life applications, to ensure unprecedented reliability and value.

Acknowledgments: We extend our sincere gratitude to all the authors who contributed their valuable research to this Special Issue. We must also express our heartfelt thanks to all the reviewers who dedicated their time and effort to provide expert feedback. Finally, we thank the editorial team of *Batteries* for their significant support throughout the organization and publication process. We hope that this collection of papers will serve as a valuable reference and source of inspiration for scholars and engineers in the field of lithium-ion batteries.

Conflicts of Interest: The author declares no conflicts of interest.

List of Contributions:

1. Savu, V.-I.; Brace, C.; Engel, G.; Didcock, N.; Wilson, P.; Kural, E.; Zhang, N. Linear Regression-Based Procedures for Extraction of Li-Ion Battery Equivalent Circuit Model Parameters. *Batteries* **2024**, *10*, 343. https://doi.org/10.3390/batteries10100343.
2. Chen, H.; Zhang, W.; Zhang, C.; Sun, B.; Yang, S.; Chen, D. Diffusion-Equation-Based Electrical Modeling for High-Power Lithium Titanium Oxide Batteries. *Batteries* **2024**, *10*, 238. https://doi.org/10.3390/batteries10070238.
3. Ayayda, M.; Benger, R.; Reichrath, T.; Kasturia, K.; Klink, J.; Hauer, I. Modeling Thermal Runaway Mechanisms and Pressure Dynamics in Prismatic Lithium-Ion Batteries. *Batteries* **2024**, *10*, 435. https://doi.org/10.3390/batteries10120435.
4. Lee, M.-H.; Choi, S.-M.; Kim, K.-H.; You, H.-S.; Kim, S.-J.; Rho, D.-S. An Evaluation Modeling Study of Thermal Runaway in Li-Ion Batteries Based on Operation Environments in an Energy Storage System. *Batteries* **2024**, *10*, 332. https://doi.org/10.3390/batteries10090332.
5. Fan, T.-E.; Chen, F.; Lei, H.-R.; Tang, X.; Feng, F. Fault Diagnosis for Lithium-Ion Battery Pack Based on Relative Entropy and State of Charge Estimation. *Batteries* **2024**, *10*, 217. https://doi.org/10.3390/batteries10070217.
6. Yao, J.; Gao, Q.; Gao, T.; Jiang, B.; Powell, K.M. A Physics–Guided Machine Learning Approach for Capacity Fading Mechanism Detection and Fading Rate Prediction Using Early Cycle Data. *Batteries* **2024**, *10*, 283. https://doi.org/10.3390/batteries10080283.
7. Liu, C.; Wang, S.; Ma, Z.; Guo, S.; Qin, Y. A Multi-Encoder BHTP Autoencoder for Robust Lithium Battery SOH Prediction under Small-Sample Scenarios. *Batteries* **2025**, *11*, 180. https://doi.org/10.3390/batteries11050180.
8. Kalk, A.; Leuthner, L.; Kupper, C.; Hiller, M. An Aging-Optimized State-of-Charge-Controlled Multi-Stage Constant Current (MCC) Fast Charging Algorithm for Commercial Li-Ion Battery Based on Three-Electrode Measurements. *Batteries* **2024**, *10*, 267. https://doi.org/10.3390/batteries10080267.

References

1. Chen, J.; Zhou, Z.; Zhou, Z.; Wang, X.; Liaw, B. Impact of Battery Cell Imbalance on Electric Vehicle Range. *Green Energy Intell. Transp.* **2022**, *1*, 100025. [CrossRef]
2. Fang, D.; Wu, W.; Li, J.; Yuan, W.; Liu, T.; Dai, C.; Wang, Z.; Zhao, M. Performance simulation method and state of health estimation for lithium-ion batteries based on aging-effect coupling model. *Green Energy Intell. Transp.* **2023**, *2*, 100082. [CrossRef]
3. Zhu, Y.; Shang, Y.; Gu, X.; Tao, X.; Li, X.; Fu, X.; Cheng, Z. Multi-Level Early Warning of Thermal Runaway Based on Internal Pressure-Temperature Fusion for Lithium-Ion Batteries. *Green Energy Intell. Transp.* **2025**, 100368. [CrossRef]
4. Chen, W.; Han, X.; Pan, Y.; Yuan, Y.; Kong, X.; Liu, L.; Sun, Y.; Shen, W.; Xiong, R. Defects in Lithium-Ion Batteries: From Origins to Safety Risks. *Green Energy Intell. Transp.* **2025**, *4*, 100235. [CrossRef]
5. Li, S.; Zhang, C.; Du, J.; Cong, X.; Zhang, L.; Jiang, Y.; Wang, L. Fault Diagnosis for Lithium-Ion Batteries in Electric Vehicles Based on Signal Decomposition and Two-Dimensional Feature Clustering. *Green Energy Intell. Transp.* **2022**, *1*, 100009. [CrossRef]

6. Lu, D.; Ren, F.; Li, C.; Ming, H.; Cui, N. Fusing Coupled Degradation Mechanisms with Machine Learning: A Multi-Fidelity Framework for Lithium-Ion Battery Lifespan Prediction. *Green Energy Intell. Transp.* **2025**, 100367. [CrossRef]
7. Jia, Y.; Luo, G.; Zhang, Y. Development of Optimal Speed Trajectory Control Strategy for Electric Vehicles to Suppress Battery Aging. *Green Energy Intell. Transp.* **2022**, *1*, 100030. [CrossRef]
8. E, L.; Wang, J.; Yang, R.; Wang, C.; Li, H.; Xiong, R. A Physics-Informed Neural Network-Based Method for Predicting Degradation Trajectories and Remaining Useful Life of Supercapacitors. *Green Energy Intell. Transp.* **2025**, *4*, 100291. [CrossRef]
9. Arunan, A.; Qin, Y.; Li, X.; Yuen, C. A Federated Learning-Based Industrial Health Prognostics for Heterogeneous Edge Devices Using Matched Feature Extraction. *IEEE Trans. Autom. Sci. Eng.* **2024**, *21*, 3065–3079. [CrossRef]

batteries

MDPI

Article

Fault Diagnosis for Lithium-Ion Battery Pack Based on Relative Entropy and State of Charge Estimation

Tian-E Fan [1,2], Fan Chen [1], Hao-Ran Lei [1], Xin Tang [1] and Fei Feng [3,4,*]

1 College of Automation, Chongqing University of Posts and Telecommunications, Chongqing 400065, China; fante@cqupt.edu.cn (T.-E.F.)
2 Chongqing Key Laboratory of Complex Systems and Autonomous Control, Chongqing University of Posts and Telecommunications, Chongqing 400065, China
3 School of Automation, Chongqing University, Chongqing 400044, China
4 Key Laboratory of Complex System Safety and Control, Chongqing University, Chongqing 400044, China
* Correspondence: feifeng@cqu.edu.cn

Abstract: Timely and accurate fault diagnosis for a lithium-ion battery pack is critical to ensure its safety. However, the early fault of a battery pack is difficult to detect because of its unobvious fault effect and nonlinear time-varying characteristics. In this paper, a fault diagnosis method based on relative entropy and state of charge (SOC) estimation is proposed to detect fault in lithium-ion batteries. First, the relative entropies of the voltage, temperature and SOC of battery cells are calculated by using a sliding window, and the cumulative sum (CUSUM) test is adopted to achieve fault diagnosis and isolation. Second, the SOC estimation of the short-circuit cell is obtained, and the short-circuit resistance is estimated for a quantitative analysis of the short-circuit fault. Furthermore, the effectiveness of our method is validated by multiple fault tests in a thermally coupled electrochemical battery model. The results show that the proposed method can accurately detect different types of faults and evaluate the short-circuit fault degree by resistance estimation. The voltage/temperature sensor fault is detected at 71 s/58 s after faults have occurred, and a short-circuit fault is diagnosed at 111 s after the fault. In addition, the standard error deviation of short-circuit resistance estimation is less than 0.12 Ω/0.33 Ω for a 5 Ω/10 Ω short-circuit resistor.

Keywords: fault detection; sliding windows; relative entropy; SOC estimation; short-circuit resistance estimation

1. Introduction

To relieve the petrochemical energy crisis and environmental pollution problem, energy storage batteries characterized by electrochemical energy storage plays a crucial role in electric vehicles (EVs) and electricity systems. Among electrochemical energy storage batteries, lithium-ion batteries (LIBs) have attracted widespread attention and are regarded as the most promising batteries [1,2] owing to their outstanding characteristics of energy density, long cycle-life and environmental protection. Additionally, the comprehensive research progress in emerging electrolyte systems for ultralow-temperature lithium batteries offers promising prospects for enhancing their performance under extreme conditions, thereby facilitating the broader application of energy storage devices in cold climates [3]. Nowadays, LIBs have been widely used in electric power storage, EVs, electronic devices, electric boats and even electric aircraft. However, safety accidents caused by LIBs have occurred frequently in recent years, such as fires and explosions, which has aroused public concerns and inhibited LIBs from further development to some extent. Therefore, to prevent safety accidents involving LIBs, it is very necessary to detect and isolate the faults of lithium-ion battery packs in the early fault period.

The reasons for safety accidents with LIBs include natural battery aging and battery abuse. Battery abuse is a major factor and can be categorized into mechanical abuse, electrical abuse and thermal abuse [4]. Moreover, battery abuse will give rise to a variety of battery faults and accelerate battery degradation, even causing a thermal runaway [5,6]. The main cause of thermal runaway is internal short circuit (ISC), which can be divided into micro-short circuit (MSC) and hard-short circuit. However, because a hard-short circuit will immediately lead to thermal runaway, it is meaningless to attempt to detect a hard-short circuit because there is not enough time to take action to avoid thermal runaway [7,8]. Therefore, it is critical to detect ISC faults, especially MSC faults. However, other types of faults will also turn into ISC faults if they are not detected in a limited period. Consequently, developing a multiple-fault diagnosis and identification method is very important for lithium-ion battery packs [9].

Common battery faults include overcharge/over-discharge, connection fault, external/internal short circuit, sensor fault and inconsistency within the battery pack. A variety of methods have been proposed for the detection of one type of these faults or multi-faults in lithium-ion battery packs. The detection methods can be classified into two categories: model-based fault diagnosis methods [10–14] and non-model-based fault diagnosis methods [15–22]. Model-based methods usually establish the battery model to obtain the parameters using fault information and a set of residual signals. Then, the battery faults can be detected by comparing the residual signals with a fault threshold. For example, Xiong et al. proposed a sensor fault diagnosis method by using an unscented Kalman filter (UKF) for SOC estimation, then combining it with a coulomb counting method to detect and isolate current/voltage sensor faults [10]. Zhang et al. proposed a model-based and entropy method for multi-fault diagnosis; the terminal voltage residual and temperature residual based on an extended Kalman filter (EKF) were used to detect and isolate multi-faults in LIBs [11]. Gao et al. developed the relationship between the main parameters of the model and SOC based on a cell difference model, with EKF used for SOC estimation and recursive least squares (RLS) for MSC diagnosis [12]. These model-based methods can effectively detect and isolate battery faults, but most of these fault diagnosis methods require the offline identification of model parameters, which indicates fault diagnosis performance degradation under different operating conditions and temperatures. In addition, if all parameters of the battery pack are identified online in a short time, it will produce a large computational burden. Non-model-based diagnostic methods (including machine learning, statistical analysis and information entropy) have been extensively applied to detect faults in LIBs. Naha et al. proposed a supervised machine learning approach to classify the faults with features extracted from charging curves [15]. In addition, the entropy or correlation coefficient of terminal voltage changes significantly if a fault has occurred, which can be used for fault detection by comparing the terminal voltage in the fault with that in normal operation. Kang et al. designed the interleaved voltage topology and correlation coefficient method for multi-fault diagnosis (voltage sensor fault, connection fault and short-circuit fault) [16]. Zhao et al. proposed a fault diagnosis method based on a machine learning algorithm and a 3σ multilevel screening strategy, and the fault occurrence frequencies of terminal voltage were calculated [17]. Moreover, information entropy methods were developed, and the entropy was calculated in real time using sliding windows to detect faults in LIBs [18,19]. Chen et al. identified the internal state parameters of a faulty battery using RLS and an EKF, and the parameters were subsequently analyzed using the local outlier factor to diagnose the fault in the battery [20]. Li et al. utilized the interclass correlation coefficient method to extract fault information from the voltage by using a sliding window, and then they designed a loop that connects the first and last voltages to locate faults in battery packs [21]. Ma et al. established a fault-free voltage model by a graph neural network and conducted fault diagnosis by comparing it with the measured voltage [22]. These methods have been validated in the test set and performed well in fault diagnosis; however, the accuracy of these methods largely depends on large numbers of fault data and is susceptible to measurement noise.

In addition, because most of the existing methods focus on single type of fault diagnosis, the quantitative analysis of short-circuit fault is still relatively few; thus, in this paper, a multi-fault diagnosis method based on relative entropy and SOC estimation is proposed for a lithium-ion battery pack. The faults include voltage sensor fault, temperature sensor fault and short-circuit fault. First, the cell with measured median voltage is selected as the reference cell, and the relative entropies of the voltage and temperature of the other cells to those of the reference cell are calculated using a sliding window. Second, the relative entropies are evaluated by the cumulative sum (CUSUM) test to achieve fault diagnosis and isolation. Next, the equivalent circuit model parameters of the reference cell are identified using recursive least squares with the variable forgetting factor (VFFRLS), and the SOC of the short-circuit cell is estimated based on the EKF. Furthermore, the short-circuit resistance is estimated with RLS for the quantitative analysis of short-circuit fault.

The main contributions of this research lie in the following aspects: (1) The proposed fault diagnosis method can apply to various operating environments because it is not necessary for a battery model to be built to obtain fault residuals. Importantly, the computational burden and hardware redundancy are not increased to accomplish fault diagnosis and quantitative analysis with the proposed method. (2) The forgetting mechanism can effectively extract the fault information of the voltage and temperature of the battery pack, and CUSUM is used to evaluate the fault information, ensuring the efficiency and accuracy of fault diagnosis. Specifically, the voltage/temperature sensor fault was detected at 71 s/58 s after faults occurred under the US06 condition, and a short-circuit fault was diagnosed at 111 s after the fault with a 5 Ω short-circuit resistor under Federal Urban Driving Schedule (FUDS). The standard error deviation of short-circuit resistance estimation is less than 0.12 Ω/0.33 Ω for a 5 Ω/10 Ω short-circuit resistor under FUDS. (3) Not only can various faults (voltage/temperature sensor faults, short-circuit fault) be diagnosed in the proposed method, but quantitative analysis on short-circuit fault is performed by estimating the resistance of the short-circuit resistor to obtain the degree of short-circuit fault. The remainder of this paper is organized as follows. The multi-fault diagnosis method is introduced in Section 2. The results of the virtual battery pack with the sensor faults and short-circuit faults are analyzed and discussed in Section 3. Finally, the key conclusions are summarized in Section 4.

2. Methods

In this section, a relative entropy method is adopted to calculate the relative entropy of cells' voltage, temperature and SOC in real time, and the CUSUM is applied to achieve fault diagnosis and isolation. During the charge or discharge process, the difference in SOC between a normal cell and a fault cell can be expressed as ΔSOC, and the CUSUM of ΔSOC is represented as CUΔSOC during the charge or discharge period. When the CUΔSOC is positive in both the charging and discharging phases, it indicates that the SOC of the fault cell is consistently lower than that of the normal cell during this period. Therefore, it can be diagnosed that the fault may be an internal short-circuit fault. Conversely, if the CUΔSOC is negative during the charge phase, it can be implied that the fault cell is likely an aged cell. Furthermore, with the diagnosed short-circuit fault, we quantitatively estimate short-circuit resistance. The specific implementation process is shown in Figure 1.

2.1. Description of Relative Entropy Method

2.1.1. Relative Entropy Calculation

Relative entropy has been widely used in information theory and data mining due to its excellent performance in measuring the difference between two probabilities' distributions [23]. Assuming that $P(x)$ and $Q(x)$ are two probability distribution functions for a discrete random variable x, the relative entropy of P to Q is defined by

$$RE(P||Q) = \sum P(x) \cdot log \frac{P(x)}{Q(x)} \quad (1)$$

where the sum of $P(x)$ and $Q(x)$ are both 1. Usually, P and Q come from frequency distributions, so the relative entropy is always positive. If P and Q are identical, the relative entropy is zero. The higher the relative entropy, the lower the similarity between two probability distributions. In this work, relative entropy is used to detect battery fault information. First, for a series-connected battery pack, the cell with the median voltage is selected and regarded as the reference cell. Next, the relative entropy of the measured values (voltage, temperature and SOC) of the other cells to the reference cell is calculated. Despite the inconsistency of cells in the battery pack, this difference is not obvious proof that the battery pack can operate normally. However, if a sensor fault occurs in a battery, the value measured by the sensor will definitely deviate from that of the reference cell. Similarly, if a short-circuit fault occurs in a battery, the measured temperature and voltage will also greatly differ from that of the reference cell. Therefore, to obtain the fault information in real time, it is necessary to calculate the relative entropy of temperature, voltage and SOC for each cell, adopting the sliding window at each sampling moment. Here, taking the voltage of a battery pack as an example, the raw voltage matrix of n cells within sampling time $t1$ to $t2$ is represented as

$$U = [U_1, U_2, \ldots U_i, \ldots U_n] = \begin{bmatrix} u_1^{t1} & u_2^{t1} & \cdots & u_n^{t1} \\ \vdots & \vdots & \ddots & \vdots \\ u_1^{t2} & u_2^{t2} & \cdots & u_n^{t2} \end{bmatrix} \tag{2}$$

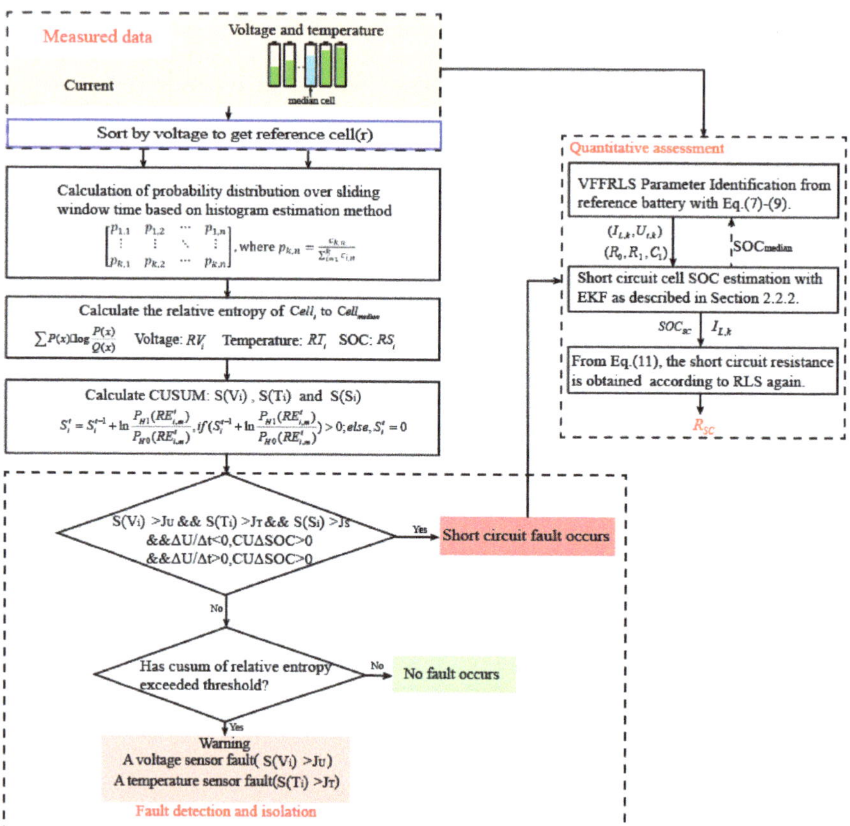

Figure 1. The schematic process of the multi–fault diagnosis method.

The voltage of the reference cell during this period can be expressed as $U_m = \left[u_m^{t1} \ldots, u_m^{t2}\right]^T$. Next, in sampling time $t1$ to $t2$, a matrix C is defined as

$$C = \begin{bmatrix} c_{1,1} & c_{1,2} & \cdots & c_{1,n} \\ \vdots & \vdots & \ddots & \vdots \\ c_{k,1} & c_{k,2} & \cdots & c_{k,n} \end{bmatrix} \quad (3)$$

where k indicates the number of intervals from $t1$ to $t2$; normally, k is set to 10 [19,24]. $c_{k,n}$ refers to the number of incidences of U_n in $\left(u_{min} + (x-1)\frac{u_{max}-u_{min}}{k}, u_{min} + x\frac{u_{max}-u_{min}}{k}\right)$ during $t1$ to $t2$, u_{max} and u_{min} are the maximum and minimum values of U, respectively, and $x = 1 \cdots k$. According to C, the frequency distribution matrix P is obtained thus:

$$P = \begin{bmatrix} p_{1,1} & p_{1,2} & \cdots & p_{1,n} \\ \vdots & \vdots & \ddots & \vdots \\ p_{k,1} & p_{k,2} & \cdots & p_{k,n} \end{bmatrix}, \text{ where } p_{k,n} = \frac{c_{k,n}}{\sum_{i=1}^{k} c_{i,n}} \quad (4)$$

The relative entropy RD^{t2} of the measured temperature and voltage of all cells to that of the reference cell is defined as $RD^{t2} = \left[RE_{1,m}^{t2}, RE_{2,m}^{t2} \cdots, RE_{n,m}^{t2}\right]$. Like the Shannon entropy, the sliding window length has a great influence on the entropy value: neither too small nor too large can reveal the battery fault information; here, the sliding window size is chosen as 100 [19,24].

2.1.2. Relative Entropy Evaluation

In order to accurately detect faults in a battery and reduce the risk of false/missed alarms, a CUSUM test is applied to evaluate the relative entropy of carrying fault information; the details can be referred to [11,25]. Fault diagnosis can be regarded as a binary hypothesis problem; based on the likelihood ratio principle, the CUSUM of the log-likelihood ratio of the sequence can be formulated as

$$S_i^t = \begin{cases} S_i^{t-1} + ln\frac{P_{H1}\left(RE_{i,m}^t\right)}{P_{H0}\left(RE_{i,m}^t\right)}, & if\ S_i^{t-1} + ln\frac{P_{H1}\left(RE_{i,m}^t\right)}{P_{H0}\left(RE_{i,m}^t\right)} > 0 \\ 0, & if\ S_i^{t-1} + ln\frac{P_{H1}\left(RE_{i,m}^t\right)}{P_{H0}\left(RE_{i,m}^t\right)} < 0 \end{cases} \quad (5)$$

where P_{H1} and P_{H0} are the probability density functions of the sequence under hypotheses H1 (fault) and H0 (no fault), respectively, and the probability distribution is viewed as Gaussian distribution. $RE_{i,m}^t$ represents the relative entropy of the measured voltage or temperature of a cell i at sampling point t. If a fault occurs, the CUSUM of log-likelihood ratios S_i^t will increase continually to exceed the alarm threshold of voltage J_U (J_T for temperature). The alarm threshold has been calibrated by extensive experimental or simulation tests.

2.2. *Quantitative Assessment Method of Short-Circuit Fault*

To evaluate the severity of a battery's short-circuit fault, it involves estimating the short-circuit current or resistance by quantitative analysis. If a short-circuit fault occurs, the real discharge electricity of a cell during discharge is larger than that calculated by coulombic counting, and the real charge electricity is smaller than that calculated by coulombic counting during charging. The battery model without/with short-circuit resistance is shown in Figure 2a/Figure 2b, respectively. In this work, firstly, the parameters (R_0, R_1, C_1) of the reference cell in the equivalent circuit model in Figure 2a are identified online by using VFFRLS. Secondly, with the identification parameters, the actual SOC of the short-circuit cell in Figure 2b is estimated by the EKF method. Next, the short-circuit current in Figure 2b is obtained by comparing it with the coulombic counting method. Furthermore, the short-circuit resistance is calculated by using RLS.

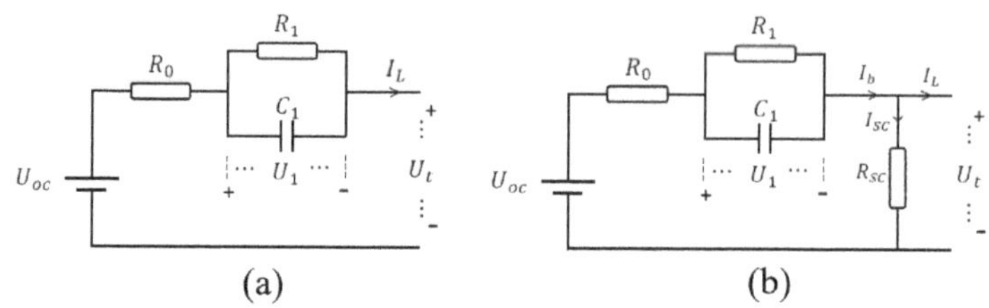

Figure 2. Structure of the equivalent circuit model: (**a**) first-order model; (**b**) first-order model with ISC.

2.2.1. Online Parameter Identification

Specifically, according to Kirchhoff's law, the dynamical electrical equations can be obtained, as shown in Equation (6). R_0 is internal resistance, R_1C_1 constitutes a network of mass transport effects and dynamic voltage performance, and U_1 describes the diffusion voltage. Open-circuit voltage U_{oc} is a nonlinear function of SOC, and U_t represents the battery terminal voltage; I_L is the load current, taken as positive for discharging and negative for charging.

$$\begin{cases} \dot{U}_1 = -\frac{1}{R_1C_1}U_1 + \frac{1}{C_1}I_L \\ U_t = U_{oc} - U_1 - I_L \cdot R_0 \end{cases} \tag{6}$$

By Laplace transformation and bilinear transformation (discretization), Equation (6) can be transformed to Equation (7), where a1, a2 and a3 are the identified parameters. With the identified parameters, the resistance and capacitance parameters in Figure 2a can be obtained as follows.

$$\begin{cases} Y_k = U_{oc,\,k} - U_{t,\,k} = \varphi_k\theta_k^T \\ \varphi_k = [Y_{k-1},\, I_{L,\,k},\, I_{L,\,k-1}] \\ \theta_k = [a_1,\, a_2,\, a_3] \\ R_0 = \frac{a_2-a_3}{1+a_1} \\ R_1 = \frac{2(a_1a_2+a_3)}{1-a_1^2} \\ R_1C_1 = \frac{1+a_1}{2(1-a_1)} \end{cases} \tag{7}$$

The process of parameter identification by the recursive least squares method with the forgetting factor is below. First, the parameter vectors θ_0 and the error covariance matrices P_0 in Equation (8) are initialized, and then the parameters (a1, a2 and a3) are adjusted adaptively according to the samples of load current and measured voltage.

$$\begin{cases} K_k = P_{k-1}\varphi_k^T\left(\lambda + \varphi_k P_{k-1}\varphi_k^T\right)^{-1} \\ P_k = \frac{1}{\lambda}[eye(3) - K_k\varphi_k]P_{k-1} \\ \theta_k^T = \theta_{k-1}^T + K_k\left(Y_k - \varphi_k\theta_{k-1}^T\right) \end{cases} \tag{8}$$

Because forgetting factor λ has great influence on recognition results, a VFFRLS algorithm is adopted to select the appropriate forgetting factor. This algorithm adaptively searches the optimal forgetting factor according to the estimation error. The forgetting factor is updated as follows.

$$\begin{cases} L(k) = -\frac{\sum_{i=k-M+1}^{k} e_i e_i^T}{M} \\ \lambda(k) = \lambda_{min} + (\lambda_{max} - \lambda_{min}) \cdot 2^{L(k)} \end{cases} \tag{9}$$

2.2.2. Short-Circuit Resistance Estimation Based on SOC

With the identified parameters of the reference cell, the SOC of the short-circuit cell can be estimated by employing a closed-loop method of EKF. By discretizing Equation (6) and combining it with the coulombic counting method, the state space equations of the battery model can be expressed as follows

$$\begin{cases} x_k = \begin{bmatrix} SOC_k \\ U_{1,\,k} \end{bmatrix} = f(x_{k-1},\, u_k) = \begin{bmatrix} SOC_{k-1} - \frac{\eta \Delta t}{C_n} \cdot I_{L,\,k} \\ exp(-\Delta t / R_1 C_1) \cdot U_{1,\,k-1} + (1 - exp(-\Delta t / R_1 C_1)) \cdot R_1 \cdot I_{L,\,k} \end{bmatrix} + w_{k-1} \\ y_k = U_{t,\,k} = g(x_k,\, u_k) = U_{oc}(SOC_k) - U_{1,\,k} - R_0 \cdot I_{L,\,k} + v_k \end{cases} \quad (10)$$

where k is the time index and Δt is the time interval. η is coulomb efficiency, assumed to be 1, and C_n is battery available capacity. x_k and y_k are the state vector and output vector at time k, respectively. f and g denote the state function and measurement function, respectively. w_{k-1} and v_k are the independent zero-mean Gaussian process noise and measurement noise with the covariance Q_k and R_k. The computational procedure of the EKF algorithm is comprehensively described in [26]. Our paper focuses on the diagnosis of internal short-circuit faults in lithium-ion batteries, so we will not elaborate further on SOC estimation here.

With the actual SOC estimation of the short-circuit battery, we can obtain the relationship between the short-circuit current and the load current, as shown in Equation (11). In consequence, the short-circuit resistance can be estimated online according to the recursive least square method.

$$\begin{cases} I_{b,\,k} = \frac{C_n \cdot (SOC_k - SOC_{k-1})}{\Delta t} \\ I_{SC,\,k} = I_{b,\,k} - I_{L,\,k} \\ U_{t,\,k} = I_{SC,\,k} \cdot R_{SC,\,k} \end{cases} \quad (11)$$

3. Results and Discussion

3.1. Data Preparation and Preprocessing

The multi-fault diagnosis of a lithium-ion battery pack was accomplished based on relative entropy and SOC estimation, including battery short-circuit fault, voltage sensor fault and temperature sensor fault. For the sensor fault tests, because sensors can experience several faults, such as deviation from the measured values to the real value, drifting, the loss of accuracy and the freezing of measured values due to inherent aging and external shocks or vibrations, an equivalent sensor fault experiment was carried out by injecting a strong noise into raw measurement data [11,16,27]. For the short-circuit fault test, destructive short-circuit experiments, such as nail penetration and squeezing, are too difficult to control, so that they are not suitable for studying the early stage of battery short-circuit. The method of cells connecting in parallel with a resistance has become an alternative experiment for battery fault tests [8,11,12,28,29]. For simulating the short-circuit fault, a resistor with a value of 5 Ω/10 Ω was selected to be connected in parallel with cell 5, representing different levels of short-circuit conditions. A dynamic thermally coupled battery model was constructed using the simulation software AutoLion-ST V2020 with a given initial value and current excitation, as shown in Figure 3 Three complete charge and discharge cycles of a battery under Federal Urban Driving Schedule (FUDS) are illustrated in Figure 3a, and the schematic diagram of different fault tests can be shown in Figure 3b. Because an electrochemical model can accurately describe battery electrochemical reactions and performance, an electrochemical–thermal coupling model in the AutoLion-ST software was developed by coupling a pseudo-two-dimensional model and the lumped mass thermal model. The parameters of the coupling model in this work were taken from [30–33]; the effectiveness of this model has been validated in [33]. The specification of the cylindrical cell was designed and is shown in Table 1.

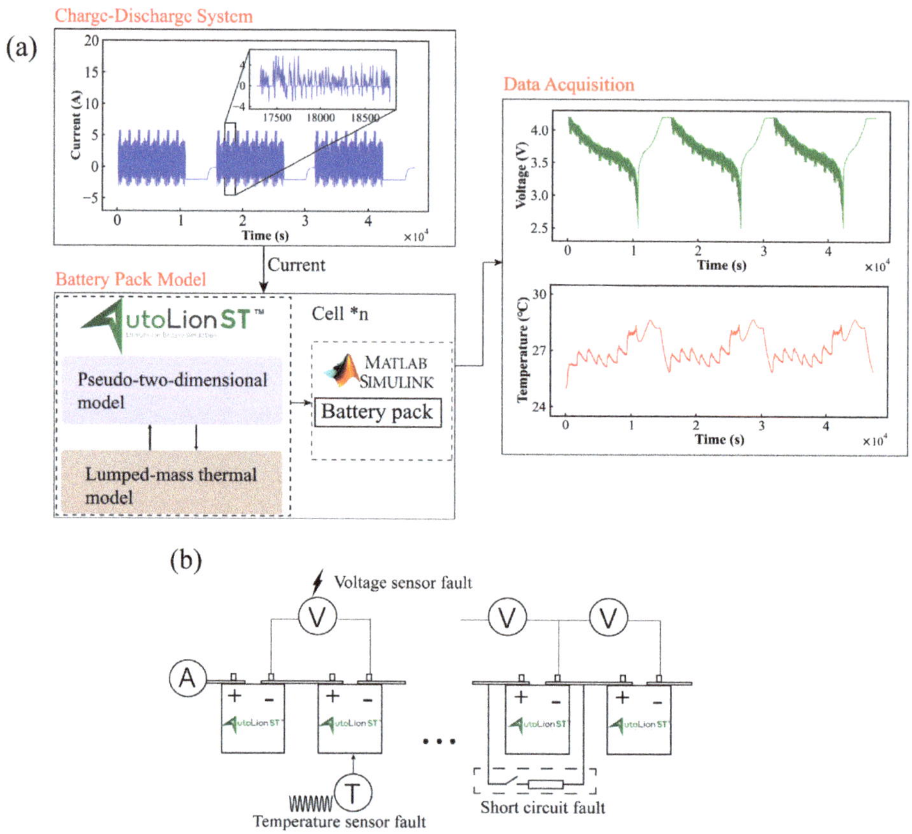

Figure 3. Schematic diagram of the virtual battery pack experiment: (**a**) Schematic diagram of the battery pack charging and discharging test; (**b**) schematic diagram of the different fault tests.

Table 1. Design parameters of cylindrical cell.

Parameters (Units)	Value
Positive electrode material	NCM111
Negative electrode material	GRAPHITE
Diameter × Height (mm)	18.6 × 65.2
Cell weight (g)	43.39
Nominal capacity (Ah)	2.15@1 C
Nominal voltage (V)	3.65
Discharge cutoff voltage (V)	2.5
Charge cutoff voltage (V)	4.2

Moreover, the SOC-OCV curve of the battery was conducted to estimate SOC, as depicted in Figure 4 A fully-charged battery was discharged to a cut-off voltage of 2.5 V, followed by charging to a cut-off voltage of 4.2 V, and the whole charge–discharge process was under a low current of 0.05 C. Two working conditions of FUDS and US06 driving schedules (US06) were considered; the specific operating condition components can be referred to in [34]. To reflect the variation of cells, different initial SOCs of the cells were selected, and the maximum inconsistency was 3%. In addition, the testing ambient temperature and the initial temperature of the battery were under 25 °C. The injected Gaussian white noise with standard deviation was 3 mV for the voltage sensor and

0.03 °C for the temperature sensor to acquire original data. A battery pack consisting of a single module with eight cells in series was built to verify the effectiveness of the proposed fault diagnosis method. Short-circuit fault and sensor fault tests were conducted under different operating conditions, where the fault parameters were selected mainly based on the experimental values in [8,11,14,35]; and the specific fault parameters for testing are listed in Table 2.

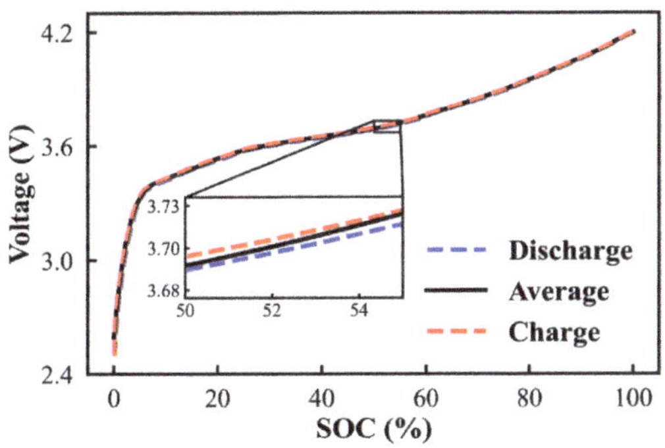

Figure 4. SOC-OCV curve under charging and discharging.

Table 2. Specific setting parameters for battery fault tests.

Fault Type	Cell No.	Fault Description	Fault Level	Trigger Time
Short-circuit	5	Resistance (Ω)	5/10	4000 s
Voltage Sensor	1	Drift (V)	0.02/0.05	5000 s
Temperature Sensor	6	Drift (°C)	0.2/0.5	5000 s

3.2. The Relative Entropy Calculation at Normal Condition

The relative entropy of voltage and temperature is used as fault information, and the CUSUM method is adopted for fault evaluation in the proposed multi-fault diagnosis method. Therefore, to obtain the statistical distribution of two hypotheses and an alarm threshold, a large number of tests were performed under both normal and fault conditions to compromise between missed alarms and false alarms. For the consideration of calculation efficiency and detection sensitivity, the sliding window was chosen as 100, and J_U and J_T were set to 150, which is more than two times their maximum CUSUM under normal conditions. To test the diagnosis effect of the proposed method on normal batteries, a normal experiment was conducted under FUDS conditions. The voltage/temperature of each cell and the relative entropies of the voltage/temperature of each cell to the reference cell in the battery pack are shown in Figure S1 and Figure S2, respectively. Here, the inconsistency in SOC in the battery pack was set to 3%, which is within a reasonable range to ensure the normal operation of the battery pack. In Figure S1, one can observe that the voltages of individual cells have certain differences during the discharge period, but the calculated voltage relative entropy fluctuates within a certain range. In addition, because of the inconsistency in all cells, the temperature difference reached a maximum at the end of discharge in Figure S2. Moreover, the relative entropies of the voltage and temperature of each cell do not deviate significantly; as can be seen from the zoomed-in figure, even the inconsistency and measurement noise were added to the battery pack.

3.3. The Results of Fault Diagnosis under FUDS Conditions

3.3.1. The Results of Sensor Faults Diagnosis

When only one sensor (either voltage or temperature) shows abnormal data, we attribute it to sensor malfunction. It is only when both sensors on the same battery simultaneously exhibit abnormal readings that we attribute the fault to the battery itself.

Therefore, upon detecting sensor faults, we do not attempt to correct the data from these faulty sensors. Instead, we discard the data generated by these faulty sensors and issue a warning. The sensor fault test was carried out by injecting a large amount noise into the original measurement data. The results of the voltage sensor fault on cell 1 and the temperature sensor fault on cell 6 under FUDS are depicted in Figure 5a,c, respectively. Because most of the sensor faults were drifting, here noise was used to simulate the fault. The noise was injected at time of 5000 s. From Figure 5a,c, one can observe that before the sensor fault happened, the measured voltage of cell 1 and the measured temperature of cell 6 fluctuated within a small noise range, just like the other cells. With noise disturbances of 0.02 V for voltage and 0.2 °C for temperature to the measurement signal, the sensor accuracy was apparently distorted, but it is difficult to detect this using the conventional threshold methods. Furthermore, the relative entropies of the voltage and temperature of each cell to the reference cell were calculated in sensor fault conditions, as shown in Figure 5b,d. From Figure 5b, it can be seen that the voltage relative entropy of cell 1 gradually deviates from that of the other cells after 5000 s. Similarly, in Figure 5d, the temperature relative entropy of cell 6 follows the same trend as the voltage relative entropy in Figure 5b. By comparing the relative entropy of cells at normal and fault, it can be indicated that the relative entropies of normal cells are not always kept at the minimum but fluctuate in a small range. Furthermore, the CUSUM method was applied to reduce the impact on sensor fault diagnosis; the result is shown in Figure 6. One can see that when the CUSUM of relative entropy exceeded a preset threshold, the voltage sensor fault was detected at 93 s after voltage disturbance was added to cell 1 and the temperature sensor fault was detected at 65 s after temperature disturbance was added to cell 6, which indicates that the sensor fault can be detected and located in a short time using our proposed method.

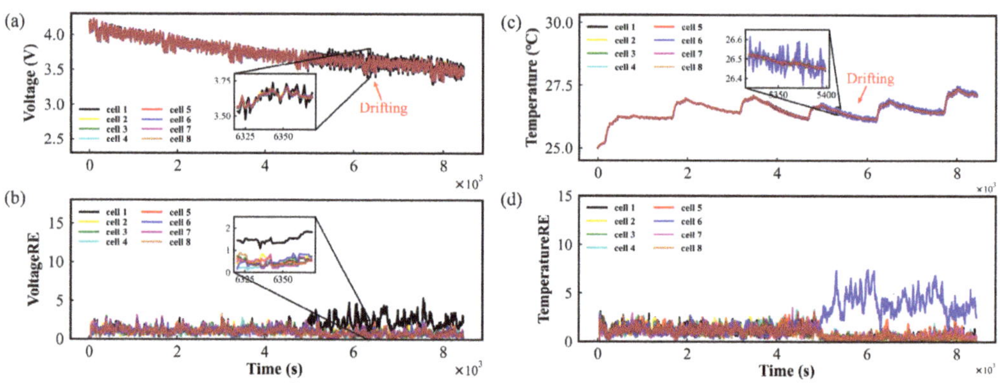

Figure 5. Results of sensor faults under FUDS conditions: (**a**) Voltage curve of each cell; (**b**) Voltage relative entropy curve of each cell to that of the reference cell; (**c**) Temperature curve of each cell; (**d**) Temperature relative entropy curve of each cell to that of the reference cell.

3.3.2. The Results of Short-Circuit Fault Diagnosis

A short-circuit fault test was conducted by connecting a resistor (5 Ω) in parallel with cell 5 under FUDS conditions; the results have been shown in Figure S3. It can be seen that the voltage of cell 5 is significantly lower and the temperature of cell 5 higher than those of the other cells after the short-circuit fault occurred. Because of the short-circuit fault on cell 5, the power consumption of the short-circuit battery cell was accelerated and reached the cut-off voltage quickly, which increased the gap between the short-circuit battery cell and the other cells to the maximum at the end of discharge. To avoid over-discharging in the short-circuit test, the battery pack was discharged until at 10–20% of its SOC. With a resistor (5 Ω) connecting it in parallel with cell 5, the voltage profile and temperature profile of all cells is depicted in Figure S3a and Figure S3c, respectively. To detect the short-circuit fault of the battery pack, the relative entropy of the voltage and temperature of all cells was calculated as fault information, as shown in Figure S3b,d. One can see that the relative entropies of the voltage and temperature of cell 5 deviated from the other cells after the short-circuit fault occurred. Furthermore, the CUSUM of the voltage and temperature relative entropy of all cells was calculated, as shown in Figure S4a,b, respectively. One can see that the CUSUM of the voltage and temperature relative entropy of cell 5 exceeded the predetermined thresholds at 4109 s and 4111 s, respectively, which indicates that a short-circuit fault of cell 5 was detected, but the severity of the short-circuit is not available.

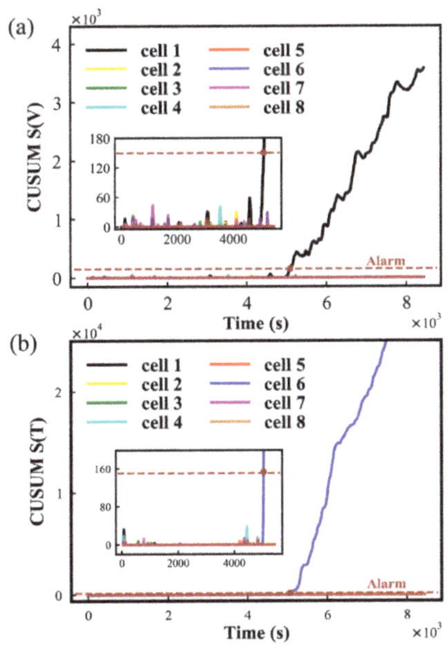

Figure 6. Results of sensor fault diagnosis under FUDS conditions: (**a**) CUSUM of voltage relative entropy; (**b**) CUSUM of temperature relative entropy.

To analyze the severity of the short-circuit of the battery pack, a quantitative evaluation was performed using SOC estimation. The SOC estimation of the reference cell (cell 4) was obtained by the EKF based on online identification parameters, as shown in Figure S5a. Here, the SOC-AH curve represents the SOC estimated using the coulombic counting method. After a resistor of 5 Ω connected with cell 5 in parallel at the time of 4000 s (when the short-circuit fault occurred), one can see that the SOC of cell 5 dropped rapidly and deviated from that of the reference cell due to the continuous power consumption by the short-circuit resistor. During this period, the SOC of a normal cell remains higher than that of a short-circuited cell, implying that the ΔSOC is consistently positive throughout the discharge phase. Moreover, the resistance of the short-circuit fault was estimated using RLS, as shown in Figure S5b. It can be observed that the mean absolute error of resistance estimation is 0.17 Ω and the standard deviation is 0.12 Ω, indicating an accurate estimation of short-circuit resistance with a small margin of error.

Furthermore, to compare short-circuit faults of different degrees, the voltage relative entropy and temperature relative entropy of all cells were obtained in the short-circuit fault of connecting a 10 Ω resistor with cell 5 in parallel, as shown in Figure S6a,b, respectively. It can be clearly observed that the deviation of the voltage relative entropy and temperature relative entropy of cell 5 was small because of the slight short-circuit fault. Usually, the smaller the resistance of connecting with cell 5 in parallel, the more serious the short-circuit fault. In addition, the CUSUM of the voltage relative entropy and temperature relative entropy of all cells were calculated to detect the short-circuit fault; the results are depicted in Figure S7. It can be seen that the CUSUM of the voltage relative entropy of cell 5 exceeded the predetermined thresholds at 4101 s and that the CUSUM of temperature relative entropy exceeded at 4200 s. By comparing the results with the short-circuit fault of connecting a 5 Ω resistor with cell 5 in parallel, one can notice that it takes longer to detect a short-circuit fault in connecting a 10 Ω resistor. Moreover, the SOC estimations of the short-circuit cell and the reference cell are shown in Figure S8a. Similarly, one can observe that the SOC of cell 5 drops rapidly and deviates from that of the reference cell at 4000 s (when the short-circuit fault occurred), which indicates that the more severe the short-circuit is, the faster the SOC drops. The resistance of the short-circuit fault was estimated and is shown in Figure S8b. The mean absolute error of resistance estimation is 0.42 Ω, and standard deviation is 0.33 Ω, indicating the effectiveness of the proposed quantitative evaluation method.

3.4. Method Comparison under US06 Condition

To further verify the performance of the proposed method, a variety of fault tests, including a short-circuit, a voltage sensor fault with 0.05 V noise and a temperature sensor fault with 0.5 °C noise, were performed to compare the proposed method with a correlation coefficient-based [36] method under US06 conditions. To simplify analysis, here, the short-circuit fault created by connecting a 10 Ω resistor with cell 5 in parallel was selected as an example. Firstly, the load current of the battery pack and the voltage and temperature of all cells are shown in Figure S9a–c, respectively. It can be seen that the voltage and temperature change in cell 5 significantly deviated from those of the other cells because of the short-circuit fault. Secondly, the fault diagnosis result based on the correlation coefficient method is displayed in Figure 7 In the correlation coefficient method, the sliding window was set to 200 for the consideration of diagnosis accuracy and computational efficiency. In Figure 7, the dashed lines parallel to the time axis are the appropriate fault thresholds derived from the tests, with minimum voltage and temperature thresholds of 0.94 and 0.6, respectively. The correlation coefficients of the voltage and temperature of each cell to the other cells were calculated in real time by a sliding window, and the maximum correlation coefficient of voltage/temperature was taken as the final diagnostic signal. From Figure 7a, it can be seen that the correlation coefficients of voltage can apparently detect the fault even with the inconsistency and noise interference. Meanwhile, the correlation coefficient of temperature fluctuates considerably in the presence of noise and inconsistency in Figure 7b. The voltage exceeded the threshold at 74 s after the fault occurred, while the temperature exceeded the threshold at 298 s. According to the abnormal correlation coefficient of voltage and temperature, it can be inferred that the short-circuit fault occurred at 298 s because a short-circuit fault influences changes in voltage and temperature. As a contrast, a short-circuit fault diagnosis based on the proposed method under US06 conditions is displayed in Figure S10. According to the CUSUM result of voltage relative entropy and temperature relative entropy in Figure S10c,d, a short-circuit fault can be diagnosed at 121 s after the fault has occurred; the time it takes to detect a short-circuit fault using our method is faster than that of the correlation coefficient method (298 s). Finally, the quantitative analysis of the short-circuit fault using SOC estimation is shown in Figure S11. The resistance of the short-circuit fault is estimated with 0.37 Ω standard deviation and 0.51 Ω mean absolute error.

In addition, a fault diagnosis test based on the Shannon entropy method was conducted, taking the voltage of a 5 Ω short-circuit resistor as an example. As shown in Figure S12a, it is difficult to directly diagnose the fault based only on Shannon entropy. Meanwhile, the correlation coefficient of Shannon entropy (Shannon RE-CC) [28] was calculated, as shown in Figure S12b. It can be seen that the Shannon RE-CC method is very sensitive to noise interference and inconsistency effects, leading to misdiagnosis or missed diagnosis by the set threshold. Therefore, the Shannon entropy-based method was not selected for comparison.

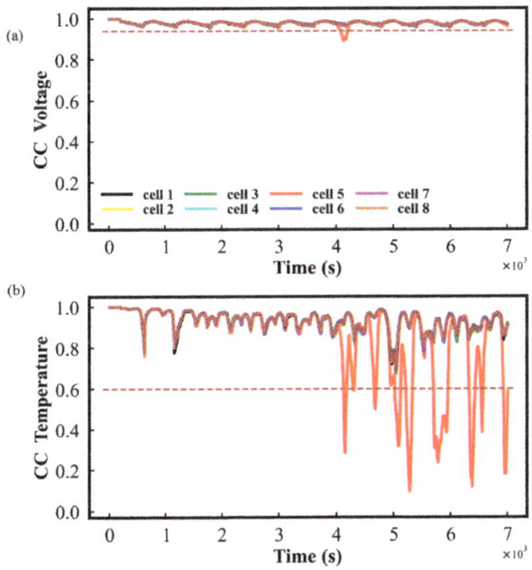

Figure 7. Short-circuit fault diagnosis results based on the correlation coefficient method: (**a**) Maximum correlation coefficient of the voltage of each cell to that of other cells; (**b**) Maximum correlation coefficient of the temperature of each cell to that of the other cells.

Moreover, different short-circuit fault degrees were tested; the fault diagnosis results are listed in Tables 3 and 4. By comparing the short-circuit fault diagnosis of our proposed method with the correlation coefficient method in different operating conditions, it can be found the proposed method performs better in short-circuit fault diagnosis for battery packs. To be specific, the proposed method had a significant advantage of diagnosis time; the smaller the short-circuit fault degree, the shorter the diagnosis time of the proposed method than that of the coefficient correlation-based diagnosis method.

Table 3. Comparison of the results of the methods proposed under FUDS and US06 with the method based on correlation coefficients [36].

Methods	FUDS		US06	
	Fault type	Detection time (s)	Fault type	Detection time (s)
Proposed method	ISC (5 Ω)	111	ISC (5 Ω)	77
	ISC (10 Ω)	200	ISC (10 Ω)	121
Correlation coefficient method	ISC (5 Ω)	281	ISC (5 Ω)	89
	ISC (10 Ω)	765	ISC (10 Ω)	298

Table 4. Resistance estimation for different short-circuit degree under different operating conditions..

Conditions	Short-Circuit Degree (Ω)	Standard Deviation Error (Ω)	Mean Absolute Error (Ω)
FUDS	5	0.12	0.17
	10	0.33	0.42
US06	5	0.13	0.21
	10	0.37	0.51

4. Conclusions

In conclusion, a fault diagnosis method based on relative entropy and SOC estimation is proposed to ensure the normal operation of a battery system, including the diagnosis of voltage sensor faults, temperature sensor faults and short-circuit faults. First, by selecting a cell with the median voltage in the battery pack to the reference cell, the relative entropies of the voltage and temperature of each cell to that of the reference one are calculated in real time according to the sliding window. Second, the relative entropy of voltage and temperature serve as fault information, and the CUSUM method is adopted for fault evaluation, indicating that the fault information of the battery is amplified by the relative entropy and the risk of false/missed alarms is reduced by the CUSUM method. In addition, VFFRLS is used for the parameter identification of the reference cell. Furthermore, by calculating the ΔSOC between a normal cell and a fault cell during both the charge and discharge phases, it becomes possible to distinguish an internal short-circuit cell from a low-capacity cell. This method can distinguish between micro-short circuit batteries and low-capacity fault batteries (or aged batteries), but it does not determine whether battery aging is due to irreversible side reactions at the positive electrode (such as oxygen release). In fact, our method cannot specifically diagnose the occurrence of irreversible side reactions (such as oxygen release) at the positive electrode of lithium batteries. Thereafter, the SOC of the short-circuit cell is estimated by an EKF, and the short-circuit resistance is estimated according to the proposed method. The effectiveness of the proposed method has been validated and compared with the correlation coefficient method under different fault types and fault degrees in FUDS and US06 with the interference of noise and inconsistency, and the results show that the proposed method performs well in both different conditions and different fault types. Meanwhile, the robustness of the proposed method was verified under the interference of noise and inconsistency in two working conditions of FUDS and US06. It should be noted that not only is the qualitative diagnosis of a short-circuit fault performed in the proposed method, but the resistance of a short-circuit resistor is estimated for an analysis of the degree of short-circuit faults. Importantly, computational effort and hardware redundancy do not increase with the proposed method, including relative entropy calculation, parameter identification and SOC estimation. In the future, the lower diagnosis time, estimation accuracy and verification of the proposed method in real-world vehicles will be the focus of our research.

Supplementary Materials: The following supporting information can be downloaded at: https://www.mdpi.com/article/10.3390/batteries10070217/s1, Figure S1. Voltage and voltage relative entropies of all cells under FUDS conditions: (a) Voltage curve of each cell; (b) Voltage relative entropy curve of each cell to that of reference cell. Figure S2. Temperature and temperature relative entropies of all cells in the battery pack under FUDS conditions: (a) Temperature curve of each cell; (b) Temperature relative entropy curve of each cell to that of reference cell. Figure S3. Short-circuit fault by connecting a resistor (5 Ω) in parallel with cell 5 under FUDS condition: (a) Voltage curve of each cell; (b) Voltage relative entropy curve of each cell to that of reference cell; (c) Temperature curve of each cell; (d) Temperature relative entropy curve of each cell to that of reference cell. Figure S4. Short-circuit fault diagnosis by connecting a resistor (5 Ω) in parallel with cell 5 under FUDS condition: (a) CUSUM result of voltage relative entropy; (b) CUSUM result of temperature relative entropy. Figure S5. Results of quantitative analysis of short-circuit fault by connecting a 5 Ω resistor with cell 5 in parallel: (a) Comparison of SOC estimation of short-circuit cell and reference cell by using the proposed method with that of coulombic counting method; (b) Result of short-circuit resistance estimation. Figure S6. Relative entropy of all cells in short-circuit fault by connecting a 10 Ω resistor with cell 5 in parallel: (a) Voltage relative entropy curve of each cell to that of reference cell; (b) Temperature relative entropy curve of each cell to that of reference cell. Figure S7. Short-circuit fault diagnosis by connecting a 10 Ω resistor with cell 5 in parallel under FUDS condition: (a) CUSUM result of voltage relative entropy; (b) CUSUM result of temperature relative entropy. Figure S8. Results of quantitative analysis of short-circuit fault by connecting a 10 Ω resistor with cell 5 in parallel: (a) Comparison of SOC estimation of short-circuit cell and reference cell by using the proposed method with that of coulombic counting method; (b) Result of short-circuit resistance estimation. Figure S9. The test of short-circuit fault with a 10 Ω resistor connecting with cell 5 in parallel: (a) Load current curve of battery pack; (b) Voltage curves of all cells; (c) Temperature curves of all cells. Figure S10. Short-circuit fault diagnosis based on the proposed method under US06 operating condition: (a) voltage relative entropy curves; (b) CUSUM result of voltage relative entropy; (c) Temperature relative entropy curves; (d) CUSUM result of temperature relative entropy. Figure S11. Results of quantitative analysis of short-circuit fault under US06: (a) Comparison of SOC estimation of short-circuit cell and reference cell by using the proposed method with that of coulombic counting method; (b) Result of short-circuit resistance estimation. Figure S12. The results of Shannon entropy and correlation coefficient of battery pack voltage under short-circuit conditions in the US06 driving cycle are presented: (a) Battery pack voltage Shannon entropy; (b) The median correlation coefficient of Shannon entropy.

Author Contributions: Conceptualization, T.-E.F.; Methodology, F.C.; Software, H.-R.L.; Validation, X.T.; Investigation, F.F. All authors have read and agreed to the published version of the manuscript.

Funding: This work is supported by the Natural Science Foundation of Chongqing (Grant Nos. cstc2021jcyj-msxmX0503, CSTB2023NSCQ-MSX0359), the Scientific and Technological Research Program of Chongqing Municipal Education Commission (Grant No. KJQN202300607), Xiaomi Foundation/Xiaomi Young Talents Program.

Data Availability Statement: The original contributions presented in the study are included in the article, further inquiries can be directed to the corresponding authors.

Conflicts of Interest: The authors declare no conflict of interest.

References

1. Xiong, R.; Sun, W.; Yu, Q.; Sun, F. Research progress, challenges and prospects of fault diagnosis on battery system of electric vehicles. *Appl. Energy* **2020**, *279*, 115855. [CrossRef]
2. Qiao, D.; Wei, X.; Fan, W.; Jiang, B.; Lai, X.; Zheng, Y.; Tang, X.; Dai, H. Toward safe carbon-neutral transportation: Battery internal short circuit diagnosis based on cloud data for electric vehicles. *Appl. Energy* **2022**, *317*, 119168. [CrossRef]
3. Yang, T.; Zheng, Y.; Liu, Y.; Luo, D.; Yu, A.; Chen, Z. Reviving low-temperature performance of lithium batteries by emerging electrolyte systems. *Renewables* **2023**, *1*, 2–20. [CrossRef]
4. Pan, Y.; Feng, X.; Zhang, M.; Han, X.; Lu, L.; Ouyang, M. Internal short circuit detection for lithium-ion battery pack with parallel-series hybrid connections. *J. Clean. Prod.* **2020**, *255*, 120277. [CrossRef]
5. Zhang, J.N.; Zhang, L.; Sun, F.C.; Wang, Z.P. An Overview on Thermal Safety Issues of Lithium-ion Batteries for Electric Vehicle Application. *IEEE Access* **2018**, *6*, 23848–23863. [CrossRef]
6. Gan, N.F.; Sun, Z.Y.; Zhang, Z.S.; Xu, S.Q.; Liu, P.; Qin, Z.A. Data-Driven Fault Diagnosis of Lithium-Ion Battery Overdischarge in Electric Vehicles. *IEEE Trans. Power Electron.* **2022**, *37*, 4575–4588. [CrossRef]
7. Seo, M.; Park, M.; Song, Y.; Kim, S.W. Online Detection of Soft Internal Short Circuit in Lithium-Ion Batteries at Various Standard Charging Ranges. *IEEE Access* **2020**, *8*, 70947–70959. [CrossRef]
8. Ouyang, M.G.; Zhang, M.X.; Feng, X.N.; Lu, L.G.; Li, J.Q.; He, X.M.; Zheng, Y.J. Internal short circuit detection for battery pack using equivalent parameter and consistency method. *J. Power Sources* **2015**, *294*, 272–283. [CrossRef]
9. Zhu, X.Q.; Wang, Z.P.; Wang, Y.T.; Wang, H.; Wang, C.; Tong, L.; Yi, M. Overcharge investigation of large format lithium-ion pouch cells with $Li(Ni_{0.6}Co_{0.2}Mn_{0.2})O_2$ cathode for electric vehicles: Thermal runaway features and safety management method. *Energy* **2019**, *169*, 868–880. [CrossRef]
10. Xiong, R.; Yu, Q.Q.; Shen, W.X.; Lin, C.; Sun, F.C. A Sensor Fault Diagnosis Method for a Lithium-Ion Battery Pack in Electric Vehicles. *IEEE Trans. Power Electron.* **2019**, *34*, 9709–9718. [CrossRef]
11. Zhang, K.; Hu, X.S.; Liu, Y.G.; Lin, X.K.; Liu, W.X. Multi-fault Detection and Isolation for Lithium-Ion Battery Systems. *IEEE Trans. Power Electron.* **2022**, *37*, 971–989. [CrossRef]
12. Gao, W.K.; Zheng, Y.J.; Ouyang, M.G.; Li, J.Q.; Lai, X.; Hu, X.S. Micro-Short-Circuit Diagnosis for Series-Connected Lithium-Ion Battery Packs Using Mean-Difference Model. *IEEE Trans. Ind. Electron.* **2019**, *66*, 2132–2142. [CrossRef]
13. Hu, J.; He, H.W.; Wei, Z.B.; Li, Y. Disturbance-Immune and Aging-Robust Internal Short Circuit Diagnostic for Lithium-Ion Battery. *IEEE Trans. Ind. Electron.* **2022**, *69*, 1988–1999. [CrossRef]
14. Liu, Z.T.; He, H.W. Sensor fault detection and isolation for a lithium-ion battery pack in electric vehicles using adaptive extended Kalman filter. *Appl. Energy* **2017**, *185*, 2033–2044. [CrossRef]
15. Naha, A.; Khandelwal, A.; Agarwal, S.; Tagade, P.; Hariharan, K.S.; Kaushik, A.; Yadu, A.; Kolake, S.M.; Han, S.; Oh, B. Internal short circuit detection in Li-ion batteries using supervised machine learning. *Sci. Rep.* **2020**, *10*, 1301. [CrossRef] [PubMed]
16. Kang, Y.Z.; Duan, B.; Zhou, Z.K.; Shang, Y.L.; Zhang, C.H. A multi-fault diagnostic method based on an interleaved voltage measurement topology for series connected battery packs. *J. Power Sources* **2019**, *417*, 132–144. [CrossRef]

17. Zhao, Y.; Liu, P.; Wang, Z.P.; Zhang, L.; Hong, J.C. Fault and defect diagnosis of battery for electric vehicles based on big data analysis methods. *Appl. Energy* **2017**, *207*, 354–362. [CrossRef]
18. Shang, Y.L.; Lu, G.P.; Kang, Y.Z.; Zhou, Z.K.; Duan, B.; Zhang, C.H. A multi-fault diagnosis method based on modified Sample Entropy for lithium-ion battery strings. *J. Power Sources* **2020**, *446*, 227275. [CrossRef]
19. Wang, Z.P.; Hong, J.C.; Liu, P. Voltage fault diagnosis and prognosis of battery systems based on entropy and Z-score for electric vehicles. *Appl. Energy* **2017**, *196*, 289–302. [CrossRef]
20. Chen, Z.H.; Xu, K.; Wei, J.W.; Dong, G.Z. Voltage fault detection for lithium-ion battery pack using local outlier factor. *Measurement* **2019**, *146*, 544–556. [CrossRef]
21. Li, X.Y.; Wang, Z.P. A novel fault diagnosis method for lithium-Ion battery packs of electric vehicles. *Measurement* **2018**, *116*, 402–411. [CrossRef]
22. Ma, G.; Xu, S.; Cheng, C. Fault detection of lithium-ion battery packs with a graph-based method. *J. Energy Storage* **2021**, *43*, 103209. [CrossRef]
23. Sun, Z.; Wang, Z.; Liu, P.; Zhang, Z.; Wang, S.; Dorrell, D.G. Relative Entropy based Lithium-ion Battery Pack Short Circuit Detection for Electric Vehicle. In Proceedings of the 2020 IEEE Energy Conversion Congress and Exposition (ECCE), Detroit, MI, USA, 11–15 October 2020; pp. 5061–5067. [CrossRef]
24. Yao, L.; Wang, Z.P.; Ma, J. Fault detection of the connection of lithium-ion power batteries based on entropy for electric vehicles. *J. Power Sources* **2015**, *293*, 548–561. [CrossRef]
25. Blanke, M.; Kinnaert, M.; Lunze, J.; Staroswiecki, M. *Diagnosis and Fault-Tolerant Control*, 3rd ed.; Springer: Berlin, Germany, 2016.
26. Zheng, Y.; Shen, A.; Han, X.; Ouyang, M. Quantitative short circuit identification for single lithium-ion cell applications based on charge and discharge capacity estimation. *J. Power Sources* **2022**, *517*, 230716. [CrossRef]
27. Kang, Y.Z.; Duan, B.; Zhou, Z.K.; Shang, Y.L.; Zhang, C.H. Online multi-fault detection and diagnosis for battery packs in electric vehicles. *Appl. Energy* **2020**, *259*, 114170. [CrossRef]
28. Qiu, Y.S.; Cao, W.J.; Peng, P.; Jiang, F.M. A novel entropy-based fault diagnosis and inconsistency evaluation approach for lithium-ion battery energy storage systems. *J. Energy Storage* **2021**, *41*, 102852. [CrossRef]
29. Xu, J.; Wang, H.T.; Shi, H.; Mei, X.S. Multi-scale short circuit resistance estimation method for series connected battery strings. *Energy* **2020**, *202*, 117647. [CrossRef]
30. Gu, W.B.; Wang, C.Y. Thermal and electrochemical coupled modeling of a lithium-ion cell in lithium batteries. *ECS Proc.* **2000**, *99*, 748–762.
31. Srinivasan, V.; Wang, C.Y. Analysis of electrochemical and thermal behavior of Li-ion cells. *J. Electrochem. Soc.* **2003**, *150*, A98–A106. [CrossRef]
32. Smith, K.; Wang, C.Y. Power and thermal characterization of a lithium-ion battery pack for hybrid-electric vehicles. *J. Power Sources* **2006**, *160*, 662–673. [CrossRef]
33. Fang, W.F.; Kwon, O.J.; Wang, C.Y. Electrochemical-thermal modeling of automotive Li-ion batteries and experimental validation using a three-electrode cell. *Int. J. Energy Res.* **2010**, *34*, 107–115. [CrossRef]
34. USABC. *Electric Vehicle Battery Test Procedures Manual. Revision 2, United States*; U.S. Department of Energy Office of Scientific and Technical Information: Oak Ridge, TN, USA, 1996. [CrossRef]
35. Feng, X.N.; Pan, Y.; He, X.M.; Wang, L.; Ouyang, M.G. Detecting the internal short circuit in large-format lithium-ion battery using model-based fault-diagnosis algorithm. *J. Energy Storage* **2018**, *18*, 26–39. [CrossRef]
36. Xia, B.; Shang, Y.; Nguyen, T.; Mi, C. A correlation based fault detection method for short circuits in battery packs. *J. Power Sources* **2017**, *337*, 1–10. [CrossRef]

batteries

MDPI

Article

Diffusion-Equation-Based Electrical Modeling for High-Power Lithium Titanium Oxide Batteries

Haoze Chen, Weige Zhang *, Caiping Zhang, Bingxiang Sun, Sijia Yang and Dinghong Chen

National Active Distribution Network Technology Research Center (NANTEC), Beijing Jiaotong University, Beijing 100044, China; hzchen@bjtu.edu.cn (H.C.); zhangcaiping@bjtu.edu.cn (C.Z.); bxsun@bjtu.edu.cn (B.S.); sjyang@bjtu.edu.cn (S.Y.); 17117408@bjtu.edu.cn (D.C.)

* Correspondence: wgzhang@bjtu.edu.cn

Abstract: Lithium titanium oxide (LTO) batteries offer superior performance compared to graphite-based anodes in terms of rapid charge/discharge capability and chemical stability, making them promising candidates for fast-charging and power-assist vehicle applications. However, commonly used battery models often struggle to accurately describe the current–voltage characteristics of LTO batteries, particularly before the charge/discharge cutoff conditions. In this work, a novel electrical model based on the solid-phase diffusion equation is proposed to capture the unique electrochemical phenomena arising from the diffusion mismatch between the positive and negative electrodes in high-power LTO batteries. The robustness of the proposed model is evaluated under various loading conditions, including constant current and dynamic current tests, and the results are compared against experimental data. The experimental results for LTO batteries exhibit remarkable alignment with the model estimation, demonstrating a maximum voltage error below 3%.

Keywords: battery model; lithium titanium oxide (LTO) batteries; rate characteristics

1. Introduction

Lithium-ion batteries have become ubiquitous in a wide range of electrical devices and systems, including telephones, electric vehicles, and renewable energy generation, owing to their high power density, high energy density, and excellent reliability [1–4]. To meet the diverse power and energy requirements of different applications, lithium-ion batteries can be categorized into two broad types: high-power designs and high-energy designs. Lithium titanium oxide (LTO) batteries utilizing $Li_4Ti_5O_{12}$ as the anode material have exhibited remarkable battery performance, encompassing excellent rate capabilities and chemical stability. Consequently, LTO batteries have emerged as a leading candidate for fast-charging and power-assist vehicle applications [5,6]. LTO batteries demonstrate pronounced advantages over conventional lead–acid batteries for starting or regenerative braking applications, which is primarily attributed to their superior power density and significantly extended service life [7–10]. LTO batteries have been used in a wide range of applications, including all-electric buses and high-speed rail trains. Highly accurate battery models are essential for ensuring safe battery operation and enhancing battery management systems [11–14].

In recent decades, various battery models have been reported to address the critical requirements of diverse scenarios. These models can be broadly categorized into three groups: electrochemical models, analytical models, and circuit models [15–18]. Electrochemical modeling is based on electrochemical equations and thermodynamic principles and represents the chemical reaction processes inside the battery numerically through partial differential equations of electrode and electrolyte kinetics [16,18]. Analytical models are founded upon the same fundamental principles as electrochemical models, with a simplification of the computational equations [11]. However, these models involve complex nonlinear differential equations and numerous unknown variables, increasing their complexity and making them challenging to integrate into battery management systems.

Circuit models, on the other hand, can capture a battery's current–voltage characteristics through a combination of electrical components such as voltage sources, resistors, and capacitors [17,19,20]. These models have a simpler structure and fewer unknown variables compared to the other two model types and can be easily incorporated into the control model of a battery-powered system. For example, Low et al. [21] proposed an improved model consisting of two resistance–capacitance (RC) parallel networks, which can predict the behavior of lithium iron phosphate (LFP) batteries with sufficient accuracy. Hu et al. [17] compared twelve circuit models using the same data, finding that a single RC parallel network model is more suitable for lithium–nickel–manganese–cobalt oxide batteries, while a first-order RC model with a single-state hysteresis is more suitable for LFP cells. Philipp et al. [22] provide a comprehensive analysis of modeling techniques for high-power LTO batteries, noting that accurately modeling battery performance at the high current rate (C-rate), extreme temperatures, and state-of-charge (SOC) boundaries are the main challenges. In addition, they conclude that second- or third-order RC equivalent circuits are best suited as LTO battery models. Many studies have also reported modeling LTO batteries with equivalent circuits, but the performance of these models is poor at high C-rates [23–25].

To improve the feasibility of high-power battery models, researchers have developed several enhanced circuit models. SOC estimation at different C-rates has been achieved using a normalization method based on the definition of the rate factor [26]. More comprehensively, Zhang et al. proposed an integrated approach combining a circuit model and a Rakhmatov diffusion model, which is capable of capturing the recovery effect [27]. However, this improved model has a complex structure that can be difficult to configure. In addition, Kim et al. used a kinetic model rather than a diffusion model to represent the rate–capacity effect to enhance the model's adaptability to high C-rates [28]. It is worth noting that the above approaches rarely discuss the variation in model parameters with current flow rate and direction, although they are sufficiently accurate from a quantitative perspective.

Lam et al. proposed an empirical formulation using curve fitting to describe the current dependence of the model parameters [29]. However, the generalization of this approach is limited due to the lack of theoretical derivation and the fact that the model validation was only achieved at rates below 2 C.

Liu et al. modeled LTO batteries in high-power applications based on the Butler–Volmer equation with a correction for the electrochemical polarization in the model [30]. Chen et al. modeled LTO batteries at different temperatures by correcting the open-circuit voltage and ohmic resistance based on the Nernst and Arrhenius equations [31]. However, these models have poor accuracy at SOC boundaries such as 0–20% SOC or 80–100% SOC. The main reason for this is caused by the deviation of the material surface SOC from the cell SOC due to the diffusion process [32].

There are several studies that have considered the effect of the diffusion process and made corrections based on diffusion equations to improve the model accuracy [27,28,32]. However, the difference in diffusion ability of positive and negative electrodes and the effect at different C-rates are not considered.

This work presents a novel electrical model for high-power LTO batteries based on the diffusion equation. The model accurately simulates the diffusive polarization of the electrodes under high C-rate operation scenarios. The accuracy and reliability of the proposed battery model across the full SOC range are validated by testing the lithium titanate battery at various C-rates and dynamic operating conditions, demonstrating a maximum error within 3%.

The remainder of this paper is structured as follows. Section 2 presents the related work, encompassing commonly utilized electrical circuit models and derivations of the diffusion equation. Section 3 details the experimental setup and procedures. The modeling process of the proposed model is outlined in Section 4. Section 5 outlines the steps for parameter extraction. Model verification results are discussed in Section 6. Finally, Section 7 concludes the work.

2. Related Work

2.1. Equivalent Circuit Model

The equivalent circuit model is a widely adopted approach for simplifying the complex electrochemical dynamics of batteries. This modeling technique leverages the electrical properties of circuit elements, such as resistors, capacitors, and voltage sources, to represent the underlying electrochemical processes within the battery. A commonly used variant is the second-order equivalent circuit model, as illustrated in Figure 1.

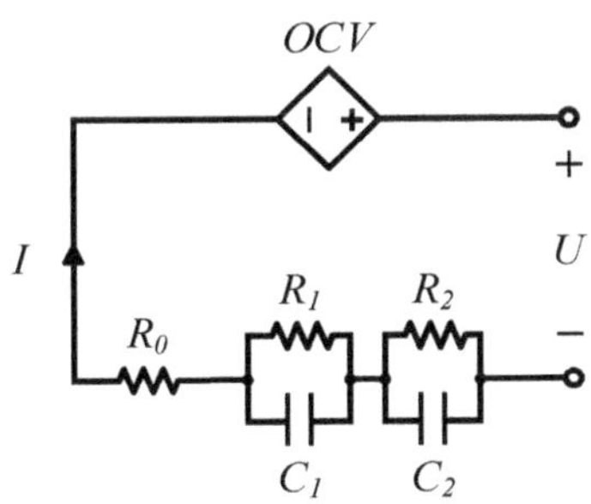

Figure 1. Schematic diagram of the second-order equivalent circuit model.

The model consists of a controlled voltage source that is primarily governed by the battery's SOC, a series resistance to account for the energy losses in the electrodes, electrolyte, and other cell components, and two parallel RC networks. These RC branches are used to simulate the electrochemical polarization and concentration polarization phenomena, respectively, which are fundamental to the battery's transient behavior.

In practice, the values of the resistances and capacitances within this equivalent circuit model are dependent on the battery's SOC and differ between charging and discharging operations. This adaptability allows the model to capture the complex and nonlinear electrochemical dynamics of the battery. However, conventional equivalent circuit models may be insufficient to accurately characterize the battery's performance under high-power applications, due to their limited ability to represent the intricate internal reaction mechanisms.

2.2. Solid-Phase Diffusion Equation

To better simulate the voltage response in high-power application scenarios, this work analyzes the effect of different polarization processes on the battery voltage. The analysis is based on the results of electrochemical impedance spectroscopy (EIS) from our previous study [33], in conjunction with analytical methods from the literature [34].

As shown in Figure 2, the various polarization losses are separated using curves simulated with parameters identified from the EIS data. Among the polarization effects, diffusion polarization has the greatest impact on the voltage, followed by ohmic polarization and interfacial polarization. The voltage losses for ohmic and interfacial polarization are fully developed within a few seconds. Therefore, the prolonged voltage loss is primarily caused by diffusion polarization. Additionally, while the interfacial polarization differs for the positive electrode (PE) and negative electrode (NE) at different SOCs, the combined effect of these two polarizations remains relatively constant. In contrast, the effects of diffusive polarization vary significantly at different SOCs.

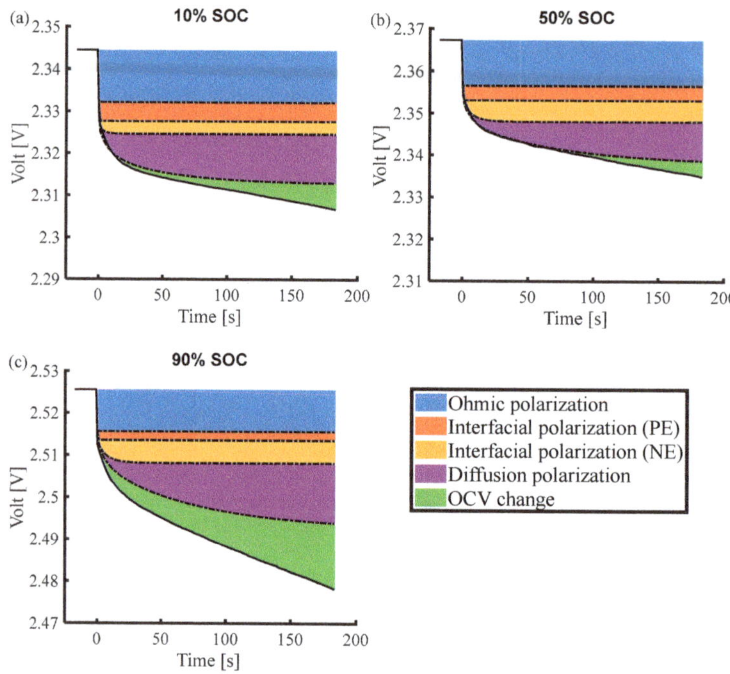

Figure 2. Polarization voltage simulation during an 180 s constant-current discharge pulse at (**a**) 10% SOC, (**b**) 50% SOC, and (**c**) 90% SOC.

The results shown in Figure 2 indicate that diffusion polarization has a greater impact than interfacial polarization on the LTO battery. Specifically, the voltage drop caused by diffusion polarization is 1.07 to 1.75 times that caused by interfacial polarization. This finding highlights the urgent need for methods that can effectively characterize the diffusive polarization resulting from concentration changes within the cell.

During the establishment of the diffusion process in the spherical electrode particles, the diffusion of lithium ions within the material follows Fick's law [35], which can be expressed as follows:

$$\frac{\partial c_s(r,t)}{\partial t} = \frac{D}{r^2}\frac{\partial}{\partial r}\left[r^2 \cdot \frac{\partial c_s(r,t)}{\partial r}\right] \tag{1}$$

where c_s is the solid-phase lithium-ion concentration, r is the distance from the center of the spherical particles, and D is the diffusion coefficient. Additionally, the following boundary conditions must be satisfied:

$$\frac{\partial c_s}{\partial r}\Big|_{r=0} = 0, \tag{2}$$

$$\frac{\partial c_s}{\partial r}\Big|_{r=R_s} = -\frac{i(t)}{nFD} \tag{3}$$

where n is the number of charges carried by a single charged particle and F is Faraday's constant. However, it seems impractical to directly obtain the concentration difference by integration, as Equation (1) cannot be solved without a known initial concentration value, which is typically unavailable. Fortunately, the initial concentration value does not affect the concentration difference between the material concentration and the average concentration. Wang and Srinivasan [36] proposed an empirical equation to describe the evolution of the concentration gradient within a spherical particle under constant current, which provides a valuable tool for characterizing the diffusive polarization in LTO cells.

$$c_s(r,t) = c_{avg}(t) + \frac{i(t)r}{nFD}\left(1 - e^{-\frac{4\sqrt{Dt}}{3r}}\right) \tag{4}$$

3. Experimental Setup and Procedures

The experimental test platform is depicted in Figure 3, comprising a battery charge/discharge test system and a temperature-controlled test chamber. The battery charge/discharge test system is a multi-channel 5V-100A tester manufactured by Arbin, featuring a voltage accuracy of ±0.02% and a current accuracy of ±0.05%. The temperature-controlled thermal chamber was provided by GIANT FORCE CO., with a temperature range of −60 °C to 100 °C and a temperature resolution of 0.1 °C. In this study, the thermal chamber was kept at a constant temperature of 25 °C.

Figure 3. Experimental setup for battery tests.

The basic parameters of the LTO battery investigated for this study are summarized in Table 1. Compared to lithium-ion batteries with graphite negative electrodes, LTO batteries are preferred for high-power applications in electric vehicles due to their superior rate capabilities. Accordingly, the reference rate for the maximum current of the battery is provided in Table 1.

Table 1. The basic parameters of LTO batteries investigated.

Battery Parameters	Characteristics
Nominal capacity	25 Ah
Voltage range	1.8~2.8 V
Max. charge current	8 C (200 A)
Max. discharge current	12 C (300 A)
Cathode material	$LiCoO_2$
Anode material	$Li_4Ti_5O_{12}$

To evaluate the battery's performance under different charge/discharge rates, a series of constant-current charge/discharge tests were conducted. The battery was first charged or discharged at a 1 C rate until the cutoff voltage was reached. Subsequently, the battery was then charged and discharged in the opposite direction at five different current rates: 1 C, 2 C, 4 C, 6 C, and 8 C. This experimental protocol allowed for the assessment of the battery's performance as a function of the applied current rates.

For internal resistance characterization, the test conditions were based on the Japanese electric vehicle standard (JEVS). The procedure involved the following steps between 10% and 90% SOC at 10% SOC intervals:

(1) 1 C charging for 60 s;
(2) 0.1 C discharging until the discharged capacity equals the charging capacity of the previous step;
(3) Resting for 10 min;
(4) 1 C discharging for 60 s;
(5) 0.1 C charging until the charging capacity equals the charging capacity of the previous step;
(6) Repeat steps (1) to (5) after replacing 1 C with 2 C, 4 C, 6 C, and 8 C, respectively.

Additionally, the battery was subjected to dynamic stress test (DST) and federal urban dynamic schedule (FUDS) profiles to provide a comprehensive dataset for model validation. Typical current curves for the DST and FUDS tests are shown in Figure 4.

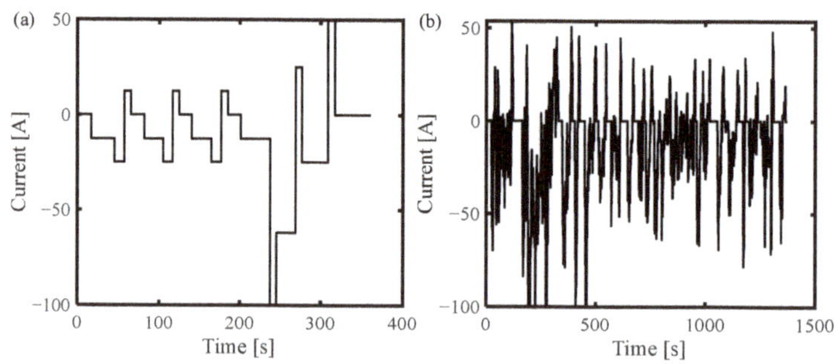

Figure 4. Current profiles for (**a**) the DST and (**b**) FUDS test.

4. Diffusion Equation Based Electrical Model

Figure 5 shows the diffusion-equation-based electrical model. It employs two controlled voltage sources to simulate the positive and negative open-circuit potential (OCP), respectively, and uniquely embeds a simplified form of the diffusion equation to simulate diffusion polarization. In addition, parallel branches of R_0 and R_1, C_1 are used to simulate ohmic and interfacial polarization, respectively. As a result, the proposed model is sufficient to characterize the integrated battery behavior, especially for lithium titanate batteries applied to high-rate applications.

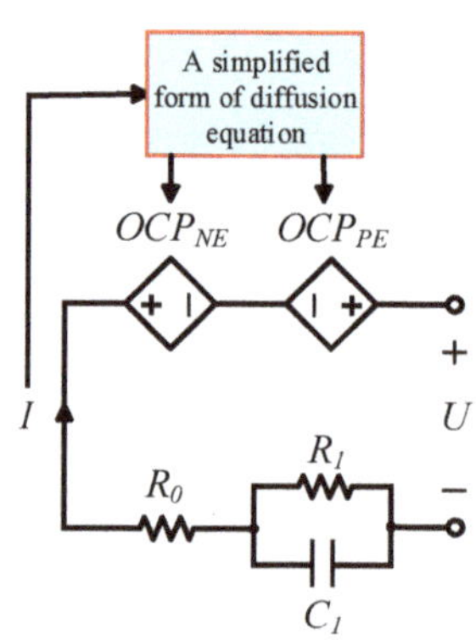

Figure 5. Schematic of the proposed electrical model.

Assuming that the battery is discharged with current I from the equilibrium state at the moment t_0, the proposed model is described as

$$U = OCP_{PE}[SOL_{PE}(t)] - OCP_{NE}[SOL_{NE}(t)] - R_0 I - R_1 I\left(1 - e^{-\frac{t}{R_1 C_1}}\right) \quad (5)$$

where OCP_{PE} and OCP_{NE} are functions of positive and negative electrode OCP with respect to the state of lithiation (SOL). Since only the lithium-ion concentration on the surface of the electrode material affects the OCP of the electrode, the expression of the electrode SOL can be obtained according to the diffusion Equation (4) as

$$SOL(t) = \overline{SOL}(t) + \Delta SOL[I(t)] \cdot \left(1 - e^{-\frac{\sqrt{t}}{\tau}}\right) \quad (6)$$

where $SOL(t)$ denotes the lithiation state on the surface of the electrode material; $\overline{SOL}(t)$ denotes the average lithiation state of the electrode material; $\Delta SOL[I(t)]$ denotes the shift of the lithiation state of the electrode due to diffusive polarization; and τ is the corresponding time constant. The $\overline{SOL}(t)$ can be calculated by integrating the current over time as shown in Equation (7), where Q is the capacity of the electrode.

$$\overline{SOL}(t) = \overline{SOL}(t_0) - \frac{1}{Q}\int_{t_0}^{t} I(t)dt \quad (7)$$

As derived in Section 2.2, the proposed model differs from other methods that aim to improve model accuracy by correcting the model parameters. In the case of the presented approach, none of the circuit elements (R_0, R_1, and C_1) within the model vary with the SOC of the battery. Instead, the diverse polarization behaviors exhibited by the cell at different SOC levels are all represented by the diffusion-based polarization term. Specifically, SOL(t) can be calculated using Equation (8) since the diffusion polarization is also changed at different current rates:

$$\Delta SOL[I(t)] = \lambda \cdot I(t) + \mu \quad (8)$$

where λ and μ are the parameters to be identified. The diffusion Equation (6) can be further morphed into

$$SOL(t) = \overline{SOL}(t) + [\lambda \cdot I(t) + \mu] \cdot \left(1 - e^{-\frac{\sqrt{t}}{\tau}}\right). \quad (9)$$

By directly incorporating diffusion-based polarization into the model structure, the proposed model is able to accurately capture the battery's performance characteristics across a wide range of SOC and current conditions without the need for extensive parameter tuning. The complete schematic for calculating the output voltage is shown in Figure 6. The U_0 and U_1 are the voltages induced by the current across R_0 and R_1/C_1, respectively.

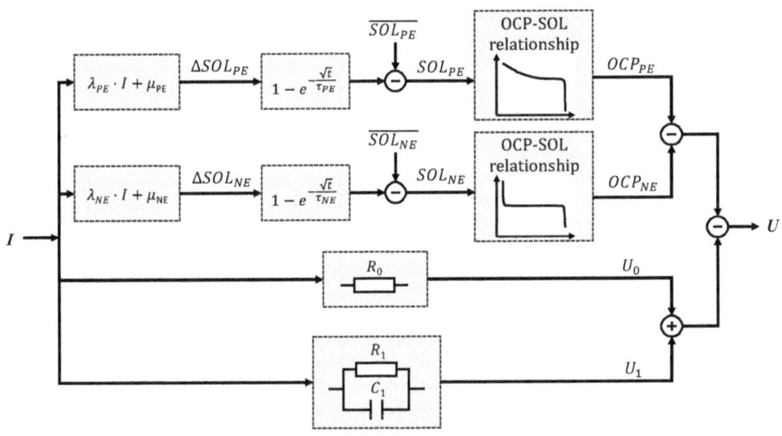

Figure 6. Schematic for calculating the output voltage.

5. Model Extraction

5.1. Full Cell OCV-SOC and Electrode OCP-SOL

As an integral part of the proposed model, the full-cell, open-circuit voltage (OCV) as a function of SOC can be obtained by averaging the battery's charge and discharge curves measured at low current rates. This approach ensures that the full-cell OCV-SOC relationship captures the underlying electrochemical equilibrium behavior of the battery. To further characterize the model, the relationship between the OCP and the SOL for the positive and negative electrodes can be determined through half-cell testing. The positive material of the half-cell is the positive and negative materials obtained from the disassembled full cell, and the negative material of the half-cell is lithium metal. Then, according to the method in the literature [37], the OCV curves of the full battery are matched using the OCP curve, and the matching relationship between the positive and negative electrodes and the full battery is obtained as shown in Figure 7.

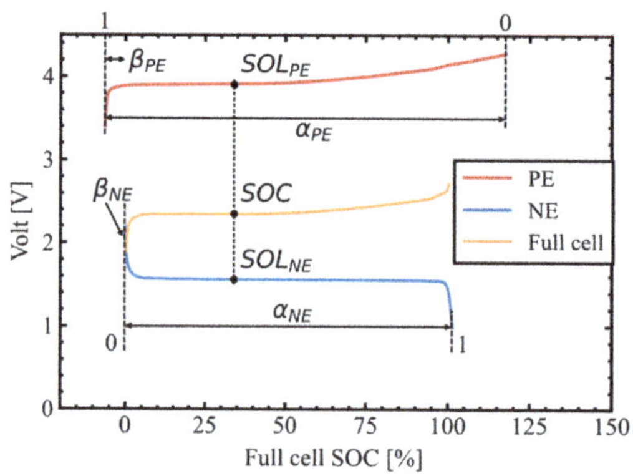

Figure 7. Matching relationship between electrode SOL and full cell SOC.

The matching process involves scaling and offsetting the electrode OCP-SOL curves to align with the OCV curve. Specifically, the scaling factors α_{PE} and α_{NE} are used to stretch or shrink the OCP curves of the positive and negative electrodes, respectively, while the offsets β_{PE} and β_{NE} are used to shift the OCP curves to lower SOC levels to match the full-cell behavior.

When $SOL = 0$, it indicates that the electrode is completely delithiated, and the electrode OCP is the highest; when $SOL = 1$, it indicates that the electrode is completely embedded in lithium, and the electrode OCP is the lowest. Finally, the relationship between the full-cell capacity (Q_{Full}) and the individual electrode capacities (Q_{PE} and Q_{NE}) can be calculated using the following equation:

$$Q_{Full} = \frac{Q_{PE}}{\alpha_{PE}} = \frac{Q_{NE}}{\alpha_{NE}}. \quad (10)$$

5.2. Parameter Identification

Depending on the parameter type, the parameter identification process can be divided into three parts:

(1) The OCV curves of the full cell were reconstructed using the OCP data of the half-cells based on the methodology presented in the literature [33,38], to obtain the correspondence parameters between the electrode and the full cell (α_{PE}, α_{NE}, β_{PE}, and β_{NE}), as well as the ohmic resistance (R_0). These parameters and Equation (10) were then used to calculate the capacity of the electrode (Q_{PE}, Q_{NE}).

(2) Based on the three-electrode battery configuration, the battery was subjected to constant-current charge/discharge experiments at different C-rates. The parameters (λ, μ) in the diffusion Equation (9) were identified based on the difference between

the diffusion polarization of the positive and negative electrodes. In addition, the parameter R_1 was determined by ohmic and interfacial polarization.

(3) Based on the JEVS experiments with different SOCs, the parameters (C_1, τ) were obtained by least-squares fitting of different pulses according to the different magnitudes of the time constant. The final parameters were obtained by averaging the parameters at different SOCs.

To investigate the SOL shift that occurs at the positive and negative electrodes due to the diffusion polarization under different current rates, this study employed a three-electrode cell configuration. As shown in Figure 8, the positive electrode is $LiCoO_2$, the negative electrode is $Li_4Ti_5O_{12}$, and a lithium metal electrode serves as the reference electrode. The inclusion of the reference electrode allowed for a clear distinction between the voltages of the positive and negative electrodes. Specifically, the voltage of the positive electrode with respect to the reference electrode is denoted as U_{PE}, while the voltage of the negative electrode with respect to the reference electrode is denoted as U_{NE}. The voltage of the full battery can be expressed as the sum of these two electrode potentials, as shown in the following equation:

$$U = U_{PE} - U_{NE}. \tag{11}$$

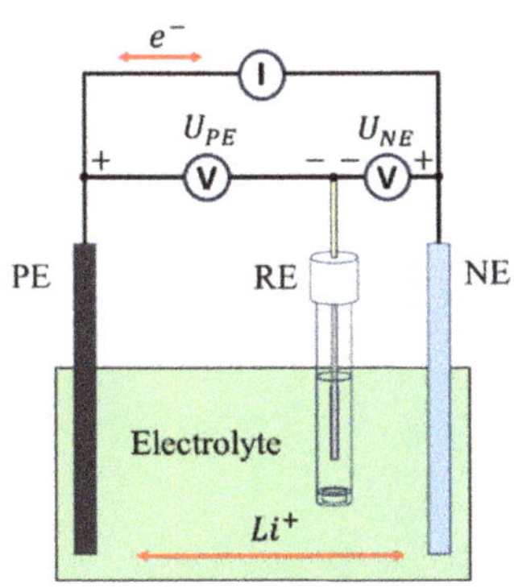

Figure 8. Schematic principle of three-electrode LTO battery.

After assembling the three-electrode lithium titanate battery, charging and discharging experiments were conducted at different current rates to highlight the distinct effects of positive and negative electrode polarization on the full-cell voltage. Two representative current conditions were selected for this analysis: the OCV test under 0.1 C (2.5A), which minimizes the influence of polarization on the electrodes; and a high-rate 8 C (200A) test, which emphasizes the impact of polarization on the electrodes. Through these experiments, the performance differences between the positive and negative electrodes under varying polarization levels could be more clearly analyzed. The experimental setup utilized three test channels: one channel measured and controlled the full-cell voltage during charging and discharging, while the other two channels independently monitored the voltages between the positive electrode and the reference electrode and between the negative electrode and the reference electrode, respectively.

Figure 9 illustrates the correlation between the positive electrode voltage, negative electrode voltage, and full-cell voltage under various charging and discharging current conditions. As observed in the OCV test, the rapid rise in the full-cell voltage near the charging cutoff is primarily attributed to a significant drop in the negative electrode voltage, indicating that the negative electrode becomes the dominant factor in determining the voltage cutoff of the full battery under these quasi-equilibrium conditions. In contrast, under the high-rate 8 C charging condition, the rapid increase in the full-cell voltage before the cutoff is mainly driven by a rapid rise in the positive electrode voltage, suggesting that the positive electrode becomes the primary influence on the full-cell voltage cutoff. The voltage difference in the vertical direction at a steady state is caused by a combination of ohmic polarization and interface polarization. The polarization resistance R_1 can be obtained by dividing this voltage difference by the current and subtracting the ohmic resistance.

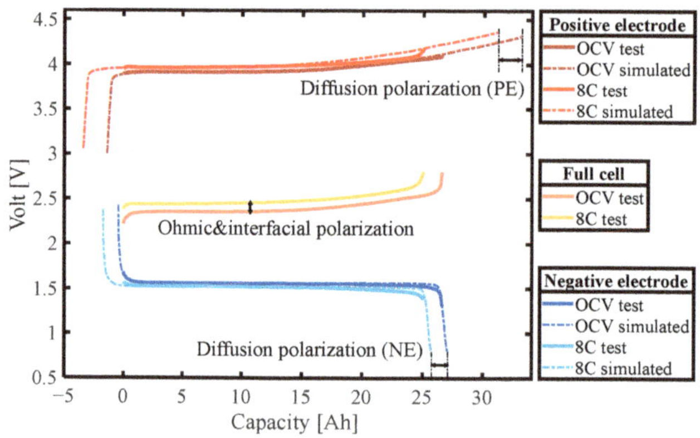

Figure 9. Relationship between voltages of positive and negative electrodes and full cell during charging.

By fitting the OCP curves of the positive and negative electrodes to the measured three-electrode voltages, the diffusion polarization at the positive and negative electrodes can be separately quantified. Additionally, the shift in the voltage direction of the full-cell curve represents the combined effects of ohmic and interfacial polarization. Similar observations can be made for the OCV and 8 C discharge tests, as shown in Figure 10. Compared to the charging process, the diffusion polarization at the positive electrode remains relatively unchanged, while the diffusion polarization at the negative electrode is significantly reduced, which may be attributable to the differences in the material properties between the positive and negative electrodes.

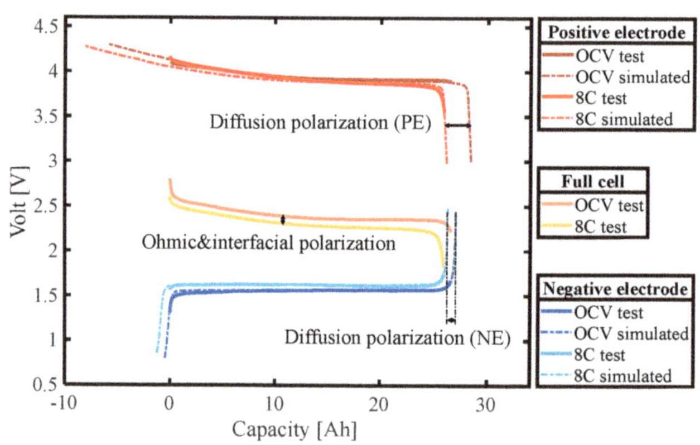

Figure 10. Relationship between voltages of positive and negative electrodes and full cell during discharging.

By matching the positive and negative electrode voltage curves obtained during constant-current charging and discharging at different current rates, the offset β^{rate} under various current rates can be obtained and used to calculate the ΔSOL^{rate} of the electrodes, where the subscript "rate" denotes the specific current rate applied during the charging and discharging processes. It is important to note that during the curve-matching process, the parameters α_{PE} and α_{NE} should be constrained to ensure consistency with the identified OCP curves. This approach ensures that the positive and negative electrodes have the same capacity. According to the matching results, the ΔSOL^{rate} of the positive and negative electrodes can be calculated using Equation (12):

$$\Delta SOL^{rate} = \beta^{OCV} - \beta^{rate} \quad (12)$$

where β^{OCV} represents the offset obtained from the OCV test. The experimentally determined ΔSOL of the positive and negative electrodes as a function of the current rate is shown in Figure 11.

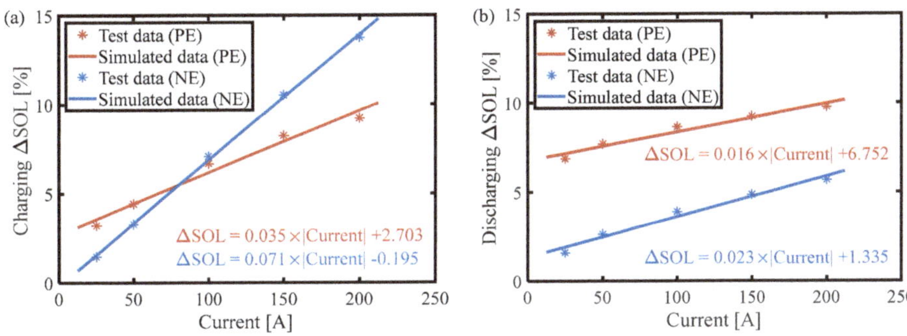

Figure 11. Positive and negative ΔSOL versus C-rates: (**a**) charging and (**b**) discharging.

After the ΔSOL has been determined, the parameters (C_1, τ) can be identified separately according to the differences in the time constants of interfacial and diffusion polarization. In general, the time constant of the diffusion process is much larger than that of the interfacial process. Therefore, the particle swarm algorithm was fitted to the model using the voltage response under the complete 60 s current pulse from the JEVS test to identify the parameter τ. It is worth noting that parameter C was set to infinity during the identification process to ignore the effect of the interfacial process. Then, the voltage response of the first 10 s of the pulse in the JEVS test was used to identify parameter C_1, which characterizes the interfacial process. All the identified model parameters at 25 °C are summarized in Table 2.

Table 2. Identification results of equivalent circuit model parameters for LTO battery.

Parameters	Current Direction	Positive Electrode	Negative Electrode	Full Cell
R_0	-	-	-	0.8 mΩ
R_1	-	-	-	0.4 mΩ
C_1	-	-	-	27,732 F
α^{OCV}	-	1.239	1.023	-
β^{OCV}	-	−0.064	−0.003	-
λ	Charge	0.009	0.018	-
	Discharge	0.004	0.006	-
μ	Charge	0.027	−0.002	-
	Discharge	0.068	0.013	-
τ	Charge	116.6 s	106.1 s	-
	Discharge	123.2 s	102.0 s	-

6. Model Verification

To validate the proposed model, constant current tests and dynamic current tests with different current rates were carried out for LTO batteries, respectively. The superiority of the proposed model was then quantitatively assessed by comparing the simulation results against the experimental data. In this study, the proposed model was implemented using MATLAB, which ensures its potential for generalization and application in high-power applications.

6.1. Galvanostatic Test at Different Rates

Figure 12 presents the validation results of the lithium titanate battery model under constant current charging conditions with varying current multiplication factors. As shown, the simulated voltage results of the proposed model closely align with the experimental test data, despite the differences in current multiplication. Specifically, when the current was 1 C, the error throughout the entire charging process was less than 1%, as shown in Figure 12a. Similarly, for an 8 C current, the error remained below 1% for the first 350 s of charging, though it exhibited an upward trend towards the end, with a maximum error of less than 3%, as shown in Figure 12b.

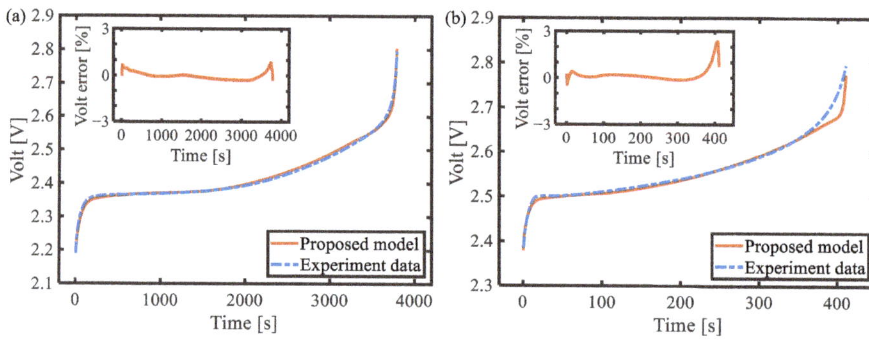

Figure 12. Results of LTO battery model under galvanostatic charging test: (**a**) 1 C rate and (**b**) 8 C rate.

On the other hand, Figure 13 compares the simulation results and experimental data for constant-current discharge under different current multiplication factors. Analogous to the charging performance, the model's errors were less than 1% for the majority of the discharge cycle. Larger errors, not exceeding 3%, were observed only at the beginning and end of discharge. These results demonstrate that the proposed model can accurately capture the constant-current discharge and charging behavior of lithium titanate batteries.

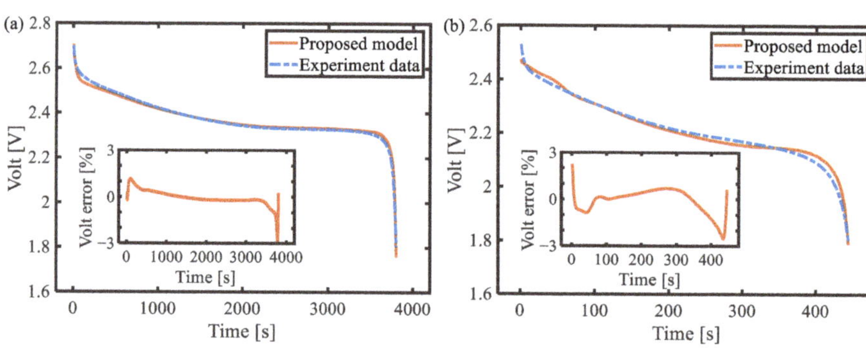

Figure 13. Results of LTO battery model under galvanostatic discharging test: (**a**) 1 C rate and (**b**) 8 C rate.

6.2. Dynamic Current Testing Using DST and FUDS Profiles

In addition to the galvanostatic validation, the proposed model was further assessed under DST and FUDS operating conditions. For the DST, as shown in Figure 14a, the error between the model-simulated voltage and the experimental data were less than 2% for the majority of the time range. This demonstrates the simulation accuracy of the model at continuous, steadily varying currents. It is worth noting that even though the error increased towards the end of the discharge period, the maximum error remained within 3%, reflecting the model's stability and reliability throughout the discharge process.

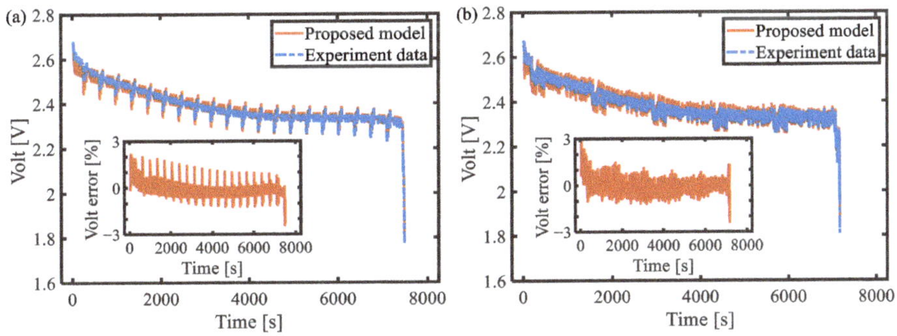

Figure 14. Results of LTO battery model under dynamic current test: (**a**) DST and (**b**) FUDS.

In addition, the battery model also performs well under more complex FUDS conditions, as shown in Figure 14b. In this case, the battery's voltage response underwent drastic oscillations due to the frequent changes in current. Nevertheless, the proposed model was still able to accurately simulate the voltage response under such dynamic conditions. Even during the most violent current fluctuations, the maximum error in the model prediction was less than 3%, proving the model's accuracy and reliability in capturing complex dynamic behavior.

6.3. Comparison against Second-Order RC Model

To better illustrate the advantages of using diffusion equation modeling, the second-order RC model and the model proposed in this paper were used to describe the performance of the battery at different conditions.

In order to compare the errors of the models under different C-rates, the errors distributed by time were transformed into errors distributed by SOC, as shown in Figure 15. The SOC was calculated based on the capacity at different C-rates, respectively. The voltage error of the second-order RC model rises continuously, exceeding 6% by the end of charge in Figure 15a. And, the error increases at a higher C-rate. Similarly, the second-order RC model also exhibits low accuracy in the galvanostatic discharging with large C-rates, as shown in Figure 15b. In contrast, the voltage error of the proposed model remains lower than 3%, demonstrating the diffusion equation's ability to better characterize the integrated battery behavior, enhancing the model's applicability for high C-rate operations.

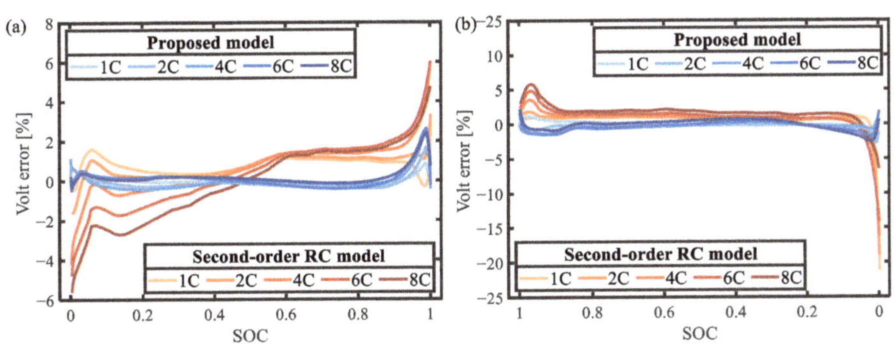

Figure 15. Comparison between the proposed model and the Thevenin model for (**a**) galvanostatic charging and (**b**) galvanostatic discharging.

Figure 16 shows the error analysis of the battery under DST and FUDS operating conditions. The voltage error of the second-order RC model increases rapidly, exceeding 18% before the discharge cutoff, while the error of the model proposed in this paper is always less than 3%. This indicates that the diffusion equation accurately captures the battery characteristics, particularly during rapid voltage changes near the charging and discharging cutoffs. The high agreement between the simulation results and experimental data for LTO batteries verifies the accuracy and reliability of the model proposed in this paper, which outperforms the second-order RC model.

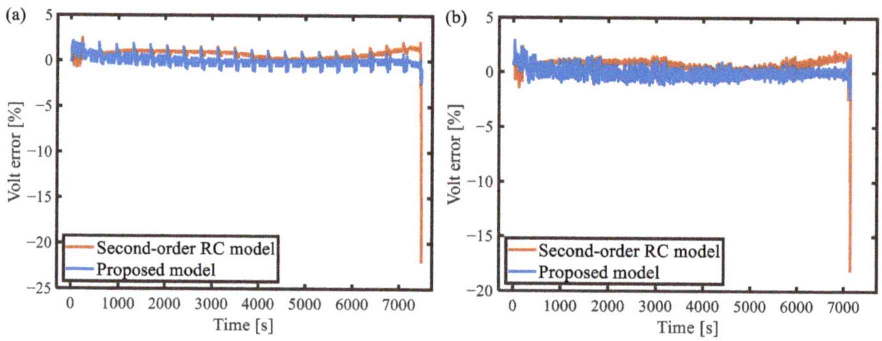

Figure 16. Comparison between the proposed model and the Thevenin model for (**a**) the DST and (**b**) the FUDS test.

7. Conclusions

In this paper, a simplified diffusion equation was used to establish an electrical model for high-power LTO batteries, and the main contributions are as follows:

(1) By comparing the polarization distribution of the battery, it was found that the effect of diffusion polarization on the LTO battery is more serious than that of interfacial polarization, and it is the diffusion polarization rather than the interfacial polarization that will change with SOC. Specifically, the effect of diffusion polarization is 1.07 to 1.75 times that of interfacial polarization.

(2) A simplified diffusion equation related to SOL that can be directly added to the circuit model was elaborately derived, which has more physical significance than the mathematical solution, and gives an ideal expression for the lithiation state of the positive and negative electrodes in the steady state.

(3) Through the three-electrode battery architecture, the degree of diffusion polarization of the positive and negative electrodes under high-power conditions was decoupled and analyzed, and the model based on the simplified diffusion equation was established with high accuracy in the full SOC range, with a maximum voltage error of less than 3%.

Compared with the second-order RC model, the model implemented in MATLAB embeds the simplified form of the diffusion equation, which reduces the model estimation error under different loading profiles. Comparative analysis of simulation and experimental data for LTO batteries shows that the proposed model has the advantage of accurately and reliably estimating the terminal voltage. In this paper, a modeling method for lithium titanate batteries is provided for the first time from the perspective of diffusion polarization and will contribute to the optimization of management strategies for battery management systems in future fast charging and electric vehicles.

Author Contributions: Conceptualization, H.C. and W.Z.; methodology, H.C.; software, H.C.; validation, H.C., C.Z. and B.S.; formal analysis, H.C.; investigation, S.Y.; resources, W.Z.; data curation, D.C.; writing—original draft preparation, H.C.; writing—review and editing, S.Y.; visualization, H.C.;

supervision, W.Z.; project administration, W.Z.; funding acquisition, W.Z., C.Z. and B.S. All authors have read and agreed to the published version of the manuscript.

Funding: This research study was funded by the National Science Fund for Distinguished Young Scholars of China (Grant No. 52222708), the Science and Technology Research and Development Plan Project of China National Railway Group (Grant No. L2022J002), the Fundamental Research Funds for the Central Universities (Grant No. 2022JBQY008), and the National Natural Science Foundation of China (Grant No. 52177206).

Data Availability Statement: The data presented in this study are available on request from the corresponding author. The data are not publicly available due to privacy.

Conflicts of Interest: The authors declare no conflicts of interest.

Nomenclature

C-rate	Current rate
DST	Dynamic stress test
EIS	Electrochemical impedance spectroscopy
FUDS	Federal urban driving schedule
JEVS	Japanese electric vehicle standard
LFP	Lithium iron phosphate
LTO	Lithium titanium oxide
NE	Negative electrode
OCP	Open-circuit potential
OCV	Open-circuit voltage
PE	Positive electrode
RC	Resistance–capacitance
SOC	State of charge
SOL	State of lithiation

References

1. Khan, F.M.N.U.; Rasul, M.G.; Sayem, A.S.M.; Mandal, N.K. Design and Optimization of Lithium-Ion Battery as an Efficient Energy Storage Device for Electric Vehicles: A Comprehensive Review. *J. Energy Storage* **2023**, *71*, 108033. [CrossRef]
2. Islam, S.; Ghazal, T.M. Lithium-Ion Battery Management System for Electric Vehicles: Constraints, Challenges, and Recommendations. *Batteries* **2023**, *9*, 152. [CrossRef]
3. Larcher, D.; Tarascon, J.M. Towards Greener and More Sustainable Batteries for Electrical Energy Storage. *Nat. Chem.* **2015**, *7*, 19–29. [CrossRef] [PubMed]
4. Park, S.; Ahn, J.; Kang, T.; Park, S.; Kim, Y.; Cho, I.; Kim, J. Review of State-of-the-Art Battery State Estimation Technologies for Battery Management Systems of Stationary Energy Storage Systems. *J. Power Electron.* **2020**, *20*, 1526–1540. [CrossRef]
5. Wegmann, R.; Döge, V.; Sauer, D.U. Assessing the Potential of a Hybrid Battery System to Reduce Battery Aging in an Electric Vehicle by Studying the Cycle Life of a Graphite|NCA High Energy and a LTO|metal Oxide High Power Battery Cell Considering Realistic Test Profiles. *Appl. Energy* **2018**, *226*, 197–212. [CrossRef]
6. Farmann, A.; Waag, W.; Sauer, D.U. Application-Specific Electrical Characterization of High Power Batteries with Lithium Titanate Anodes for Electric Vehicles. *Energy* **2016**, *112*, 294–306. [CrossRef]
7. Bank, T.; Alsheimer, L.; Löffler, N.; Uwe, D. State of Charge Dependent Degradation Effects of Lithium Titanate Oxide Batteries at Elevated Temperatures: An in-Situ and Ex-Situ Analysis. *J. Energy Storage* **2022**, *51*, 104201. [CrossRef]
8. Chahbaz, A.; Meishner, F.; Li, W.; Ünlübayir, C.; Uwe Sauer, D. Non-Invasive Identification of Calendar and Cyclic Ageing Mechanisms for Lithium-Titanate-Oxide Batteries. *Energy Storage Mater.* **2021**, *42*, 794–805. [CrossRef]
9. Yang, S.; Zhang, C.; Jiang, J.; Zhang, W.; Zhang, L.; Wang, Y. Review on State-of-Health of Lithium-Ion Batteries: Characterizations, Estimations and Applications. *J. Clean. Prod.* **2021**, *314*, 128015. [CrossRef]
10. Yang, S.; Zhang, C.; Chen, H.; Wang, J.; Chen, D.; Zhang, L.; Zhang, W.; Yang, S.; Yang, S.; Zhang, C.; et al. A Hierarchical Enhanced Data-Driven Battery Pack Capacity Estimation Framework for Real-World Operating Conditions with Fewer Labeled Data. *J. Energy Chem.* **2023**, *91*, 417–423. [CrossRef]
11. Wang, Y.; Zhang, X.; Li, K.; Zhao, G.; Chen, Z. Perspectives and Challenges for Future Lithium-Ion Battery Control and Management. *eTransportation* **2023**, *18*, 100260. [CrossRef]
12. Liu, W.; Placke, T.; Chau, K.T. Overview of Batteries and Battery Management for Electric Vehicles. *Energy Rep.* **2022**, *8*, 4058–4084. [CrossRef]
13. Yang, S.; Zhang, C.; Jiang, J.; Zhang, W.; Gao, Y.; Zhang, L. A Voltage Reconstruction Model Based on Partial Charging Curve for State-of-Health Estimation of Lithium-Ion Batteries. *J. Energy Storage* **2021**, *35*, 102271. [CrossRef]

14. Yang, S.; Zhang, C.; Jiang, J.; Zhang, W.; Chen, H.; Jiang, Y.; Sauer, D.U.; Li, W. Fast Screening of Lithium-Ion Battery Packs for Second Use with Machine Learning. *eTransportation* **2023**, *17*, 100255. [CrossRef]
15. Zhang, C.; Li, K.; McLoone, S.; Yang, Z. Battery Modelling Methods for Electric Vehicles—A Review. In Proceedings of the 2014 European Control Conference (ECC), Strasbourg, France, 24–27 June 2014; pp. 2673–2678. [CrossRef]
16. Miguel, E.; Plett, G.L.; Trimboli, M.S.; Lopetegi, I.; Oca, L.; Iraola, U.; Bekaert, E. Electrochemical Model and Sigma Point Kalman Filter Based Online Oriented Battery Model. *IEEE Access* **2021**, *9*, 98072–98090. [CrossRef]
17. Hu, X.; Li, S.; Peng, H. A Comparative Study of Equivalent Circuit Models for Li-Ion Batteries. *J. Power Sources* **2012**, *198*, 359–367. [CrossRef]
18. Kim, J.; Chun, H.; Kim, M.; Han, S.; Lee, J.W.; Lee, T.K. Effective and Practical Parameters of Electrochemical Li-Ion Battery Models for Degradation Diagnosis. *J. Energy Storage* **2021**, *42*, 103077. [CrossRef]
19. Watrin, N.; Roche, R.; Ostermann, H.; Blunier, B.; Miraoui, A. Multiphysical Lithium-Based Battery Model for Use in State-of-Charge Determination. *IEEE Trans. Veh. Technol.* **2012**, *61*, 3420–3429. [CrossRef]
20. Greenleaf, M.; Li, H.; Zheng, J.P. Modeling of LixFePO4 Cathode Li-Ion Batteries Using Linear Electrical Circuit Model. *IEEE Trans. Sustain. Energy* **2013**, *4*, 1065–1070. [CrossRef]
21. Low, W.Y.; Aziz, J.A.; Idris, N.R.N.; Saidur, R. Electrical Model to Predict Current-Voltage Behaviours of Lithium Ferro Phosphate Batteries Using a Transient Response Correction Method. *J. Power Sources* **2013**, *221*, 201–209. [CrossRef]
22. Schröer, P.; van Faassen, H.; Nemeth, T.; Kuipers, M.; Sauer, D.U. Challenges in Modeling High Power Lithium Titanate Oxide Cells in Battery Management Systems. *J. Energy Storage* **2020**, *28*, 101189. [CrossRef]
23. Madani, S.S.; Schaltz, E. An Electrical Equivalent Circuit Model of a Lithium Titanate Oxide Battery. *Batteries* **2019**, *5*, 31. [CrossRef]
24. Schröer, P.; Khoshbakht, E.; Nemeth, T.; Kuipers, M.; Zappen, H.; Sauer, D.U. Adaptive Modeling in the Frequency and Time Domain of High-Power Lithium Titanate Oxide Cells in Battery Management Systems. *J. Energy Storage* **2020**, *32*, 101966. [CrossRef]
25. Maria, E.D.; Dallapiccola, M.; Aloisio, D.; Brunaccini, G.; Sergi, F.; Moser, D.; Barchi, G. A Data-Driven Equivalent Circuit Model's Parameter Estimation Method Applied to Lithium-Titanate Battery. *J. Energy Storage* **2023**, *74*, 109497. [CrossRef]
26. Gao, L.; Liu, S.; Dougal, R.A. Dynamic Lithium-Ion Battery Model for System Simulation. *IEEE Trans. Components Packag. Technol.* **2002**, *25*, 495–505. [CrossRef]
27. Zhang, J.; Ci, S.; Sharif, H.; Alahmad, M. An Enhanced Circuit-Based Model for Single-Cell Battery. In Proceedings of the 2010 Twenty-Fifth Annual IEEE Applied Power Electronics Conference and Exposition (APEC), Palm Springs, CA, USA, 21–25 February 2010; pp. 672–675. [CrossRef]
28. Kim, T.; Qiao, W. A Hybrid Battery Model Capable of Capturing Dynamic Circuit Characteristics and Nonlinear Capacity Effects. *IEEE Trans. Energy Convers.* **2011**, *26*, 1172–1180. [CrossRef]
29. Lam, L.; Bauer, P.; Kelder, E. A Practical Circuit-Based Model for Li-Ion Battery Cells in Electric Vehicle Applications. In Proceedings of the 2011 IEEE 33rd International Telecommunications Energy Conference, Amsterdam, The Netherlands, 9–13 October 2011; pp. 1–9. [CrossRef]
30. Liu, S.; Jiang, J.; Shi, W.; Ma, Z.; Wang, L.Y.; Guo, H. Butler-Volmer-Equation-Based Electrical Model for High-Power Lithium Titanate Batteries Used in Electric Vehicles. *IEEE Trans. Ind. Electron.* **2015**, *62*, 7557–7568. [CrossRef]
31. Chen, A.; Zhang, W.; Zhang, C.; Huang, W.; Liu, S. A Temperature and Current Rate Adaptive Model for High-Power Lithium-Titanate Batteries Used in Electric Vehicles. *IEEE Trans. Ind. Electron.* **2020**, *67*, 9492–9502. [CrossRef]
32. Ouyang, M.; Liu, G.; Lu, L.; Li, J.; Han, X. Enhancing the Estimation Accuracy in Low State-of-Charge Area: A Novel Onboard Battery Model through Surface State of Charge Determination. *J. Power Sources* **2014**, *270*, 221–237. [CrossRef]
33. Chen, H.; Chahbaz, A.; Yang, S.; Zhang, W.; Sauer, D.U.; Li, W. Thermodynamic and Kinetic Degradation of LTO Batteries: Impact of Different SOC Intervals and Discharge Voltages in Electric Train Applications. *eTransportation* **2024**, *21*, 100340. [CrossRef]
34. Zhou, X.; Huang, J.; Pan, Z.; Ouyang, M. Impedance Characterization of Lithium-Ion Batteries Aging under High-Temperature Cycling: Importance of Electrolyte-Phase Diffusion. *J. Power Sources* **2019**, *426*, 216–222. [CrossRef]
35. Newman, J.; Tiedemann, W. Porous-electrode Theory with Battery Applications. *AIChE J.* **1975**, *21*, 25–41. [CrossRef]
36. Wang, C.Y.; Srinivasan, V. Computational Battery Dynamics (CBD)—Electrochemical/Thermal Coupled Modeling and Multi-Scale Modeling. *J. Power Sources* **2002**, *110*, 364–376. [CrossRef]
37. Dubarry, M.; Truchot, C.; Liaw, B.Y. Synthesize Battery Degradation Modes via a Diagnostic and Prognostic Model. *J. Power Sources* **2012**, *219*, 204–216. [CrossRef]
38. Lu, D.; Scott Trimboli, M.; Fan, G.; Zhang, R.; Plett, G.L. Implementation of a Physics-Based Model for Half-Cell Open-Circuit Potential and Full-Cell Open-Circuit Voltage Estimates: Part II. Processing Full-Cell Data. *J. Electrochem. Soc.* **2021**, *168*, 070533. [CrossRef]

batteries

MDPI

Article

An Aging-Optimized State-of-Charge-Controlled Multi-Stage Constant Current (MCC) Fast Charging Algorithm for Commercial Li-Ion Battery Based on Three-Electrode Measurements

Alexis Kalk [1,†], Lea Leuthner [2,†,*], Christian Kupper [1] and Marc Hiller [1]

[1] Institute of Electrical Engineering (ETI), Karlsruhe Institute of Technology (KIT), 76131 Karlsruhe, Germany; alexis.kalk@kit.edu (A.K.); christian.kupper@kit.edu (C.K.); marc.hiller@kit.edu (M.H.)
[2] Institute for Applied Materials (IAM), Karlsruhe Institute of Technology (KIT), 76131 Karlsruhe, Germany
* Correspondence: lea.leuthner@kit.edu
† These authors contributed equally to this work.

Abstract: This paper proposes a method that leads to a highly accurate state-of-charge dependent multi-stage constant current (MCC) charging algorithm for electric bicycle batteries to reduce the charging time without accelerating aging by avoiding Li-plating. First, the relation between the current rate, state-of-charge, and Li-plating is experimentally analyzed with the help of three-electrode measurements. Therefore, a SOC-dependent charging algorithm is proposed. Secondly, a SOC estimation algorithm based on an Extended Kalman Filter is developed in MATLAB/Simulink to conduct high accuracy SOC estimations and control precisely the charging algorithm. The results of the experiments showed that the Root Mean Square Error (RMSE) of SOC estimation is 1.08%, and the charging time from 0% to 80% SOC is reduced by 30%.

Keywords: fast charging; multi-stage constant current (MCC) charging; Li-plating; SOC; aging

1. Introduction

The global market share of the electrical vehicles has been growing exponentially in recent years. Lithium-ion battery (LIB) technology makes an important contribution to the electrification of power trains due to their high energy and power density, and long service time. Thanks to the great reduction of the battery prices in the last three decades, electric vehicle (EV) manufacturers can build high-capacity batteries in their flagship models to overcome the range anxiety [1]. However, due to the cost burden and design restrictions, this approach is not suitable for Light EVs (LEV) and small/middle class EVs, which are very crucial for the universal electrification of mobility. Besides driving range, other criteria like purchase price, lack of charging infrastructure, and charging time are very important for customer acceptance of electromobility [2].

In the last decade, the number of public chargers increased exponentially worldwide, and the share of fast chargers also grew [3]. However, charging time depends not only on the capabilities of the charger, but also on the battery properties, environmental conditions, and charging protocol. The standard charging protocol for LIB is the constant current–constant voltage (CC-CV) [4]. CC-CV is a two-stage protocol. In the CC-stage, the battery is charging at a constant current up to the upper voltage limit (e.g., 4.2 V), and then in CV-stage, charging continues by constant voltage until a predefined cut-off current (e.g., 0.05 C, where C represents the charge rate at which a battery can be fully charged or discharged in one hour) [5]. The CC-CV protocol is easy to implement, but it is not an optimal method for fast charging because of the very long CV phase. Increasing the current rate in the CC-Stage extends the charging time in CV-Stage, thus the reduction of the total charging time is limited. Moreover, high current rates can provoke lithium plating (Li-plating), i.e., deposition of metallic Li on the anode surface, which accelerates the irreversible energy and power degradation of the battery and can cause safety issues [6]. Since automotive applications demand a long battery lifetime (e.g., 10–15 years) for economic feasibility, the effect of the charging protocol on the battery service life cannot be neglected.

Several charging protocols have been proposed in the literature to optimize the trade-off between charging time, charging efficiency, and service life. Refs. [7,8] provide a detailed summary of these charging protocols. Many studies show that Li-plating is the predominant aging and failure mechanism during charging [9,10]. Dendrite growth induced by Li-plating is considered one of the most critical factors triggering battery failures [11–14], though, in fact, deposition of Li-metal should actually not be even possible in a cell under ideal conditions. It is only thermodynamically feasible if the anode surface potential falls below 0 V vs. Li/Li^+. During charging, the potential of typically used graphite anodes is in the range of approx. 1 V and 0.005 V, and does not fall below this critical limit, provided that the operating voltage window of the full-cell is below 4.2 V [15]. However, under real operating conditions, several factors can alter the half-cell potential, and can, locally or on a larger scale, lead to a drop in anode surface potential. Therefore, in practice, Li-plating is hard to prevent, and eventually occurs in small amounts in most LIB cells. It can, however, be minimized by consideration of crucial factors. Besides temperature, the upper cut-off voltage U_{MAX}, and constructive properties (cell balancing and geometry), the charging C-rate significantly affects the anode potential [9,16,17]. Therefore, charging patterns that allow more control over current are particularly suitable for the development of healthy fast charging strategies. In this aspect, the varieties of multi-stage constant current charging (MCC) protocols have been widely investigated. The MCC charging protocol basically consists of two or more CC-stages. Additionally, a CV-stage can be implemented when the cell voltage reaches the end-of-charge voltage. The shifting between stages can be based on SOC intervals or U_{MAX} [8]. The MCC charging protocol can reduce the charging time, prolong the cycle life and standing out from many other charging methods by its implementation simplicity [18–21]. To prevent Li-plating and to reduce the charging time simultaneously, selection of the optimal number of CC stages, current rates, and shifting conditions is the main challenge in the development of MCC charging. This can be handled as a combinatorial multi-criteria optimization problem, but testing every combination is not applicable. Therefore, the Taguchi method with orthogonal arrays is proposed to systematically reduce the necessary number of experiments [18,20]. Also, optimization algorithms such as particle swarm optimization (PSO) [21] and ant colony system (ACS) [22] are applied to find an optimal current pattern with different objectives, like high charging efficiency, reduced charging time, low heat generation, less capacity loss, etc. In addition to these approaches, detection of Li-plating can help to find the optimal charging current pattern. Various methods have been introduced to detect Li-plating based on cell measurements, like analysis of Coulombic efficiency, impedance, voltage, thickness, etc. [23–27]. Also, Li-plating can be directly detected by the three-electrode method. In this method, anode potential vs. Li/Li^+ is measured. A potential drop to 0 V indicates Li-plating [28]. Although this method is easy to implement, it needs a reference electrode, which is not available in commercial cells. However, experimental three-electrode cells can be built with a reference electrode using collected electrodes from the commercial cell to be investigated. In such an experimental setup, the influence of the charging rate and possibly other parameters on the potential at the negative electrode-electrolyte interface can be determined. These measured correlations can then be used for fast charging of commercial cells made of the same electrodes, but without the three-electrode setup. Several investigations show the suitability of this approach for the development of optimized charging current [9,28–30].

Our research advances the development of fast charging protocols by incorporating SOC-based thresholds derived from the three-electrode cell experiments to govern a multi-stage constant current (MCC) charging protocol for commercial cells. This SOC-based approach offers enhanced stability over conventional MCC protocols, which primarily rely on voltage-based control. This is crucial, as the voltage characteristic of LIBs can be significantly affected by both external factors, like temperature variations, and internal factors, such as electrochemical hysteresis, while the majority of SOC-based MCC algorithms in the literature are developed through expensive cell testing and lack detailed electrochemical analysis [18,20–22]. Alternatively, Epding et al. [30] introduced a SOC-based MCC charging protocol for LTO cells based on three-electrode experiments. In that work, the aging performance of the proposed protocol is validated on lab-scale experimental three-electrode cells, without extending it to the application on commercial cells. Sieg et al. [28], on the other hand, successfully applied insights from three-electrode experiments to larger-scale pouch cells, employing SOC as a critical transfer parameter from experimental to commercial cells. However, their focus remains on large-scale pouch cells designed primarily for EV applications. Their SOC estimation relies on high-accuracy laboratory current measurements (Coulomb counting), incorporating a SOC estimation error of 2% to reflect realistic conditions in selected tests.

Our study enhances this existing framework by adapting the methodology for cylindrical commercial cell formats, integrating a high-accuracy SOC estimator, and extending its relevance to NMC (Lithium Nickel Manganese Cobalt Oxides ($LiNi_xMn_yCo_{1-x-y}O_2$)) cell chemistry. The chosen commercial cell, with its 4.8 Ah nominal capacity, is well-suited for light EVs, such as electric bicycles. This not only validates the theoretical foundations of our approach, but also emphasizes its scalability for diverse industrial applications, particularly in the rapidly growing light EV sector.

The novelty of our work lies in the integration of precise SOC thresholds obtained from three-electrode experiments for MCC charging algorithm with a highly accurate SOC estimator. This synergy enables the development of scalable charging current guidelines for standard commercial cells, bridging the gap between experimental insights and their practical implementations. Our approach eliminates the need for cells with a physical third electrode during application or extensive cell testing in the development phase of the charging protocol. By doing so, it aims to reduce charging time and, through minimizing the risk of Li-plating, prevent the acceleration of aging caused by fast charging. This contributes significantly to the advancement of lithium-ion battery charging technologies.

This paper is organized as follows: First, the optimal charging pattern is designed with the help of three-electrode method. Experimental three-electrode cells are reconstructed from commercial 21700 type NMC cells. Second, an accurate SOC estimator suitable for a Battery Management System (BMS), namely an Extended Kalman Filter (EKF), is developed to ensure the proper control of the charging current on real application. Third, cell tests are performed to prove the performance of the proposed approach. Aging tests are conducted, and the proposed MCC protocol is compared with standard CC-CV charging. In the final section, the conclusions are given.

2. Materials and Methods

2.1. Electrochemical Characterization

The electrodes of NMC commercial cylindrical cells of type 21700 were analyzed electrochemically using three-electrode measurements. For this purpose, the commercial cells first were cycled for 5 cycles at standard conditions according to the manufacturer specifications (Table 1), and then discharged to U_{MIN}. This "pre-cycling" is performed to allow completion of possibly incomplete electrolyte decomposition and SEI forming processes that might manipulate the later acquired data. The cells were then opened under inert atmosphere in an argon-filled glovebox by use of appropriate tools (Figure 1a). After the extraction, the so-called jelly roll (Figure 1b) consisting of anode, cathode, and separator was carefully unrolled, and the three components were isolated from each other. Then, the electrodes were washed with dimethyl carbonate (DMC) to remove any electrolyte residues. Furthermore, the electrode coating was carefully removed from the anode and cathode current collector foil, each on one side without damaging the opposite side's coating. From the resulting one-sided coated electrodes, individual circular electrodes were punched and reassembled in three-electrode cells (Swagelok T-cell type), as shown in Figure 2.

Table 1. Cycling conditions for the analyzed cells according to manufacturer specifications.

Lower Cut-Off Voltage U_{MIN}	Upper Cut-Off Voltage U_{MAX}	Charge Mode	Discharge Mode	Temperature
2.65 V	4.2 V	CC-CV, C/2 rate	CC, 1 C rate	25 °C

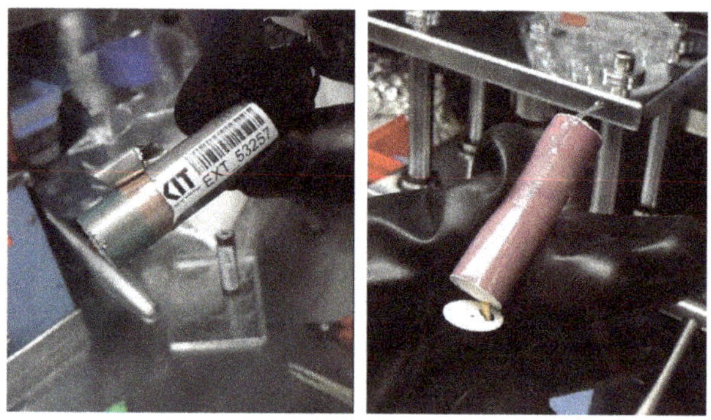

(a) (b)

Figure 1. Opening of commercial 21700 cell under inert conditions in the glovebox. (**a**) Cell opening; (**b**) jelly roll.

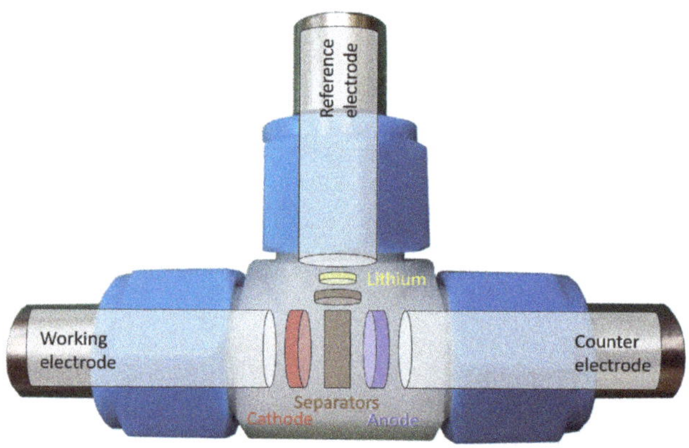

Figure 2. Swagelok T-cell with schematic design of three-electrode configuration with working electrode (WE), counter electrode (CE), and reference electrode (RE).

In the three-electrode configuration, the cell is operated at a specific full cell potential, i.e., defined by the potential difference between the working electrode (WE) and the counter electrode (CE). Ideally, the CE has a constant potential, so that the potential changes and the underlying processes at the WE can be analyzed.

However, using common LIB electrode materials (in this study, NMC as WE and graphite as CE), the potentials at both electrodes vary, simultaneously depending on the SOC and the operating conditions, which makes it hard to study the underlying processes at each electrode [31]. To observe the processes at the WE and CE in isolation, an additional reference electrode (RE) (e.g., Li | Li^+) is therefore required, which taps the voltage at the WE and CE without itself being actively involved in the electrochemical reaction. Each three-electrode cell was equipped with an extracted cathode as WE (⌀12 mm) and anode as CE (⌀13 mm) being separated by a fresh glass fiber separator (Whatman GF/D, ⌀13 mm). Lithium foil was used as reference electrode, which was additionally separated by a glass fiber separator (⌀3 mm). As an electrolyte, 450 µL of 1 M lithium hexafluorophosphate ($LiPF_6$) in a mixture of ethylene carbonate (EC) and dimethyl carbonate (DMC) at a weight ratio of 1:1 was used (LP30, Gotion Inc., Fremont, CA, USA).

The electrochemical characteristics of our experimental three-electrode cells are very similar to those of standard commercial cells. Voltage values are traditionally used as thresholds for MCC protocols due to their straightforward application. However, the voltage of LIBs is highly non-linear, influenced by SOC, temperature, and hysteresis, which can lead to inconsistencies in charge current control. Alternatively, the change in actual available capacity (Ah) offers a more linear characteristic, and is more suitable for charge current control. However, the capacity of the experimental cell differs significantly from that of commercial cells, making direct application obtained thresholds impractical. The SOC, as a normalized parameter, bridges this gap and allows the transfer of the results from the three-electrode experimental cell to the commercial cells. For this purpose, the area of coated electrode surface of the unrolled jelly roll was determined. Since the cathode fundamentally determines the amount of cyclable lithium in the system, the areal capacity [$mAh\,cm^{-2}$] was determined as a ratio of the 1 C discharge capacity and the cathode area of the commercial 21700 cell.

Then, a rate test was performed on the three-electrode cells, in which the amperage was defined according to this areal capacity. Charging and discharging took place in the full-cell voltage window of 2.65 V–4.2 V, where the discharge rate was set to C/2 and the charge rate was varied from C/10 to C/5, C/2, 1 C, and 2 C after every two cycles. By using the three-electrode setup, the anode and cathode potentials were observed individually and the critical SOC for a potential onset of Li-plating at the anode, i.e., when the anode potential falls below 0 V, can be determined for each C-rate. Eventually, the MCC protocol is normalized based on the measured three-electrode cell capacity, which allows us to scale up the charging protocol to the commercial cells.

The cells were cycled with a Biologic VMP3 battery cycler. Pre-cycling and long-term cycling of the 21,700 cells was performed by use of a BaSyTec XCTS battery cycler. The experiments were conducted at 25 °C. Future work will involve validating this method under various ambient conditions to ensure its robustness in real-world applications.

2.2. Cell Modeling and Parameter Identification

In this research, we utilize a single RC branch Thevenin equivalent circuit model (ECM) to accurately mimic the electrical characteristics of LIBs and to develop a foundational model for SOC estimation. Figure 3 visualizes the used model. It comprises the cell's open circuit voltage (U_{OCV}), the terminal voltage of the cell (U_{Cell}), and the polarization voltage (U_{RC}), along with R_0 representing the ohmic resistance, and R_1 and C_0 corresponding to the polarization resistance and capacitance, respectively. The cell current, I, is considered positive during discharge and negative during charging.

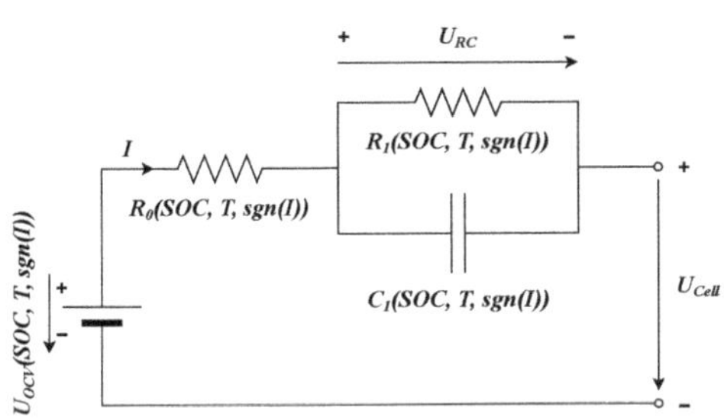

Figure 3. Thevenin ECM with 1 RC branch.

The model's mathematical representation, including Equations (1) and (2), is derived using Kirchoff's laws of current and voltage. These equations are fundamental to the model's integration to SOC estimator and simulation within the MATLAB/Simulink R2020b environment of MathWorks.

$$U_{Cell} = U_{OCV} - U_{RC} - R_0 I \tag{1}$$

$$\dot{U_{RC}} = \frac{-U_{RC}}{C_1 R_1} + \frac{I}{C_1} \tag{2}$$

The precise parametrization of these elements was conducted with a Hybrid Pulse Power Characterization (HPPC) test [32] using a BaSyTec XCTS battery cycler. To capture the thermal dependencies, the tests are carried out under seven different temperatures: −5 °C, 5 °C, 15 °C, 25 °C, 35 °C, 45 °C, and 55 °C. Additionally, to incorporate the hysteresis effect in our model, we conduct these tests in both charging and discharging directions. The HPPC tests are performed in increments of 10% SOC. The derived parameter values are then compiled into 3D lookup tables.

This extensive development and parametrization of the battery model set the stage for the subsequent chapter, where we will delve into the implementation of an Extended Kalman Filter (EKF) for precise SOC estimation. The detailed modeling and parametrization are crucial for ensuring the accuracy and efficacy of the SOC estimation process in our proposed SOC-dependent, aging-optimized fast charging algorithm.

2.3. State of Charge Estimation

The change in the SOC of a LIB can be represented as a time-dependent function, as demonstrated in Equation (3). Here, $SOC(t)$ represents the SOC at a specific time t, $SOC(t_0)$ is the initial SOC at the starting time t_0, $I(\tau)$ is the current as a function of time, and Q_n is the nominal capacity.

$$SOC(t) = SOC(t_0) - \frac{1}{Q_N} \int_{t_0}^{t} I(\tau) d\tau, \tag{3}$$

Coulomb counting is the fundamental method for SOC estimation based on Equation (3). Unfortunately, it is susceptible to errors arising from noisy current measurements and uncertainties in the initial SOC value, as well as deviations caused by charge transfer due to electrochemical side reactions. These limitations can lead to cumulative errors over time. To address these challenges, model-based approaches are employed, which offer enhanced accuracy. Consequently, in our work, we have incorporated the EKF for SOC estimation to overcome the inherent drawbacks of the Coulomb counting method and ensure more reliable and precise SOC calculations.

The EKF is a recognized adaptation of the classic Kalman filter, specifically designed for estimating states in nonlinear systems like LIBs. The EKF approach involves linearizing nonlinear systems at each time step using a first-order Taylor approximation. It is a well-established technique in battery management systems. Originally, Plett [33] pioneered the application of the EKF method for SOC estimation in LIBs. Employing a discrete-time cell model, the EKF incorporates measurement signals of current, voltage, and temperature. As a model-based recursive technique, it effectively addresses the primary challenges in SOC estimation, such as measurement noise, online estimation, and initial value problem [34]. Consequently, this has led to a significant focus among researchers on developing EKF-based SOC estimators [35–37].

Based on the Thevenin ECM as outlined in Equations (1) and (2), and the SOC as defined in Equation (3), the process and measurement equations for the EKF are formulated in the discrete-time domain as follows:

$$x_k = \begin{bmatrix} U_{RC,k} \\ SOC_k \end{bmatrix} = \underbrace{\begin{bmatrix} 1 - \frac{\Delta t}{R_1 C_1} & 0 \\ 0 & 1 \end{bmatrix}}_{A} \begin{bmatrix} U_{RC,k-1} \\ SOC_{k-1} \end{bmatrix} + \underbrace{\begin{bmatrix} \frac{-\Delta t}{C_1} \\ \frac{-\Delta t}{3600 Q_N} \end{bmatrix}}_{B} \underbrace{I_{k-1}}_{u_{k-1}} + w_{k-1}, \tag{4}$$

$$y_k = \underbrace{U_{OCV,k} - U_{RC,k} - R_0 I_k}_{h_k} + v_k, \tag{5}$$

where x_k is the state vector, w_k the process noise, y_k the cell voltage measurement vector, v_k the measurement noise, h_k the nonlinear measurement function, A the state transition matrix, B the control input matrix, u_{k-1} the input vector, Q_N the nominal capacity, k discrete-time instant, and Δt is the sampling time in seconds.

In this study, we represent estimated values using a hat symbol ($\hat{}$), and a priori predictions are indicated by a minus symbol ($^-$).

The EKF algorithm can be broken down into two main steps: prediction and update. In the prediction step, the EKF forecasts the future state $\hat{x}_k^-$ of the battery based on its current state and input variables. This involves updating the state prediction, which encapsulates the expected battery behavior, and the error covariance prediction P_k^-, which estimates the uncertainty of this prediction. Following this, the EKF proceeds to the update step, where it incorporates new measurement data to correct the predicted state. This correction is achieved by calculating the Kalman Gain K_k, which balances the prediction with new measurements, and then updating the state and error covariance estimates to more accurately reflect the observed battery behavior. Table 2 summarizes the implemented EKF algorithm. The EKF assumes that w_k and v_k are independent, zero-mean Gaussian noise processes, each characterized by their respective covariance matrices, Q and R. The linearization of the measurement function h_k is realized with Jacobian matrix C_k, as defined in Equation (6).

$$C_k = \left. \frac{\partial h(x_k, u_k)}{\partial x_k} \right|_{\hat{x}_k^-, u_k} \tag{6}$$

Table 2. Summary of prediction and update steps in Extended Kalman Filter algorithm.

1. Prediction Step	
State Prediction	$\hat{x}_k^- = A_{k-1}\hat{x}_{k-1} + B_{k-1}I_{k-1}$
Error Covariance Prediction	$P_k^- = A_{k-1}P_{k-1}A_{k-1}^T + Q$
2. Update Step	
Kalman Gain Calculation	$K_k = P_k^- C_k^T (C_k P_k^- C_k^T + R)^{-1}$
State Estimation Update	$\hat{x}_k = \hat{x}_k^- + K_k(y(k) - \hat{y}(k))$
Error Covariance Update	$P_k = (1 - K_k C_k) P_k^-$

2.4. Aging Analysis

In our study, we conducted cycling tests on three cells using the standard charging procedure, as shown in Table 1, and on two cells using the proposed MCC charging algorithm, to investigate the impact of these charging algorithms on the aging process. In both procedures, charging is extended with a CV charging phase at the end until I < C/20 and the cells are discharged with a constant current of 1 C. For all cells, a capacity test and a direct current internal resistance (RiDC) test at various SOC (10%, 30%, 50%, 70%, 90%) are conducted every 50 cycles. During the capacity test, the cells are charged using the standard CCCV charging procedure and then discharged at a constant current of 1 C until they reach the lower cut-off voltage U_{MIN}.

The RiDC tests, which can be considered a variation of the Galvanostatic Intermittent Titration Technique (GITT) profile, involve applying current pulses of 1 C in the discharge direction at each determined SOC level for 20 s, followed by a relaxation phase of 5 min. After each RiDC test, the cell is discharged to the next SOC level with a constant current of C/3, followed by a relaxation phase of 15 min. The DC internal resistances were determined by Ohm's law according to Equation (7) using the difference between the potentials at rest state before the pulse U_{t_0} and at the end of the pulse $U_{(t_0+20s)}$, caused by the applied pulse current I.

$$RiDC = \frac{U_{t_0} - U_{(t_0+20s)}}{I} \quad (7)$$

The degradation of cells is described by the State of Health (SOH). In the context of this paper, we interpreted SOH as a capacity loss and computed using Equation (8), where Q_{act} is the actual capacity of an used cell and Q_{new} is the initial reference capacity of a new cell. The reference capacity Q_{new} is determined with capacity tests with 1C discharge rate at 25 °C conducted on new cells at the start of the aging tests. The aging test was performed until the end of life of the cells, i.e., 80% SOH.

$$SOH = \frac{Q_{act}}{Q_{new}} \quad (8)$$

3. Results

3.1. Results of Electrochemical Characterization

Figure 4 shows the results of the electrochemical characterization of the three-electrode cell for determination of the SOC-dependent maximum charge rate (at 25 °C). Figure 4a shows the potentials measured at anode and cathode against the reference electrode, as well as the resulting full cell potential exemplary for the charge cycle with C/10 rate. During charging, when Li ions are extracted from the cathode and intercalate into the active material of the anode, the anode potential is reduced. In turn, the Li host material in the cathode is electrochemically oxidized and the cathode potential increases. The anode potential shows a flat plateau shape curve, which is typical for graphite anode material [15]. At the rate of C/10, it does not fall below 0 V, i.e., Li-plating is not expected. When the full cell potential reaches U_{MAX} the charge mode switches from CC to CV and the current decreases which, in turn, leads to a relaxation of overpotentials at the electrode surfaces reflected by an increase in anode potential and a decrease in cathode potential.

Figure 4b illustrates the change in anode potential relative to the SOC during charging at various C-rates, specifically C/10, C/5, C/2, 1 C, and 2 C. With higher C-rates, an increasing negative shift in anode potential can be observed, eventually leading to a drop below 0 V at C-rates $\geqq$ C/2. As a consequence, when the C rate increases, the maximum SOC that can be achieved at an anode potential of >0 V gradually becomes smaller. The increase in anode potential at higher SOC is caused by a decrease in charge current when the transition from CC- to CV-charge mode takes place. With increasing C-rate the SOC at the beginning of the CV step is decreasing, being approx. 95% at C/5, 79% at C/2, 63% at 1 C, and 29% at 2 C. Maintaining the anode potential above 0 V during charging is imperative to prevent Li-plating and to preserve the battery's operational safety. Based on these findings, we have devised a multi-stage constant current (MCC) charging profile aimed at reducing charging times while simultaneously preventing Li-plating. Figure 5 visualizes the SOC-dependent charging stages of the proposed MCC charging protocol.

Table 3 summarizes the details of the proposed MCC stages with outlining the duration required to charge through each phase. The standard charging protocol recommended by the cell manufacturer employs CCCV charging with a CC phase with a constant current rate of C/2 up to 4.2 V, and a CV phase with a cut-off current of C/30. The MCC profile provides a time-saving advantage for the lower SOC range. It completes the charge up to 80% in 67.5 min, which is about 30% faster then standard charging, which takes 96 min.

Table 3. Overview of the proposed MCC charging strategy.

SOC Range (%)	0–15	15–40	40–80	80–95	95–100
SOC Share (%)	15	25	40	15	5
C-Rate	2 C	1 C	C/2	C/5	CV
Charging Time (min.)	4.5	15	48	45	-

A full charge requires approximately 140 min, including a 38 min CV phase with standard charging, and it takes 127 min with a 15 min minute CV phase using MCC charging. The proposed MCC charging method is about 10% faster in the case of a full charge that includes a CV phase.

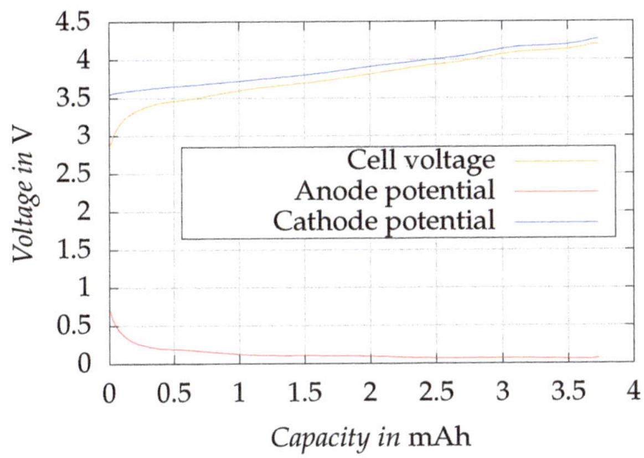

(**a**)

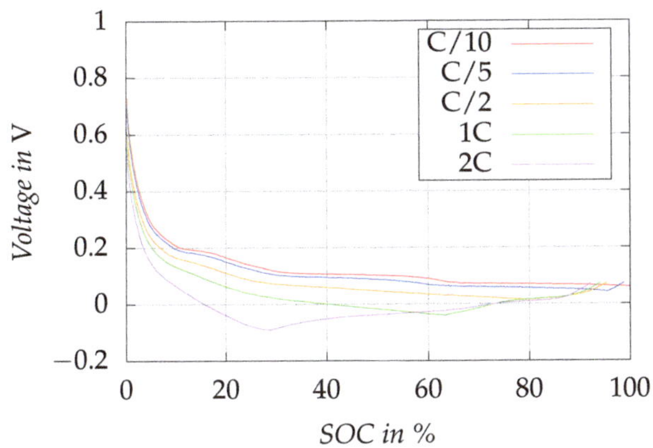

(**b**)

Figure 4. Charge curves of electrodes from commercial 21700 cell measured in three-electrode configuration. (**a**) Anode and cathode potential, as well as resulting cell potential measured during charging with C/10. (**b**) Anode potential during charging with different C-rates. The increase in anode potential at higher SOC is caused by a decrease in charge current when the transition from CC- to CV-charge mode takes place.

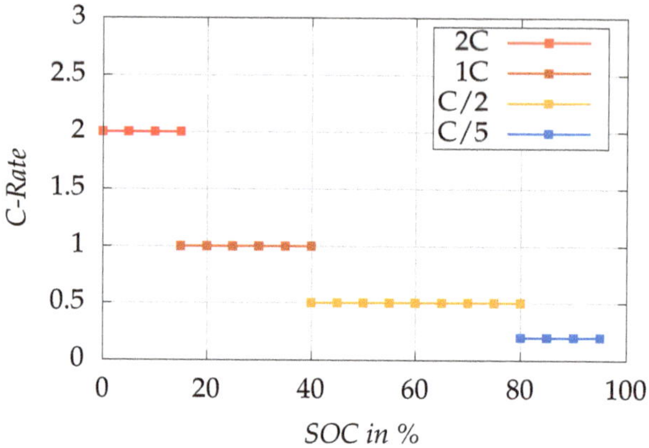

Figure 5. SOC-dependent, dynamic charging rate for commercial 21700 cell at 25 °C.

3.2. Results of Parameter Identification and Cell Modeling

The results of the HPPC tests are analyzed in Matlab. The parameters of the cell model, namely open-circuit voltage U_{OCV}, and the resistances and capacitance R_0, R_1 and C_1, are identified for various temperatures and SOC levels using Matlab's optimization function "fminsearch()".

Figure 6a illustrates U_{OCV} by charging and discharging at 25 °C. The cell's nominal capacity is specified as 4.8 A h. The influence of temperature on the cell capacity was examined under a 1 C discharge current, and the results are visualized in Figure 6b. As expected, the capacity decreases with descending temperatures. The results of the capacity test are incorporated into the cell model using temperature-dependent 2D lookup tables. Similarly, Figure 7a demonstrates the relationship of R_0 with temperature and SOC, while Figure 7b illustrates the SOC-dependent behaviors of R_0, R_1 and C_1 at 25 °C. As expected, R_0 decreases with increasing temperature. The parameter identification results indicate that the effect of SOC on the model parameters R_0, R_1 and C_1 is limited. Consequently, for simplicity, they are considered constant with respect to SOC in Equation (5).

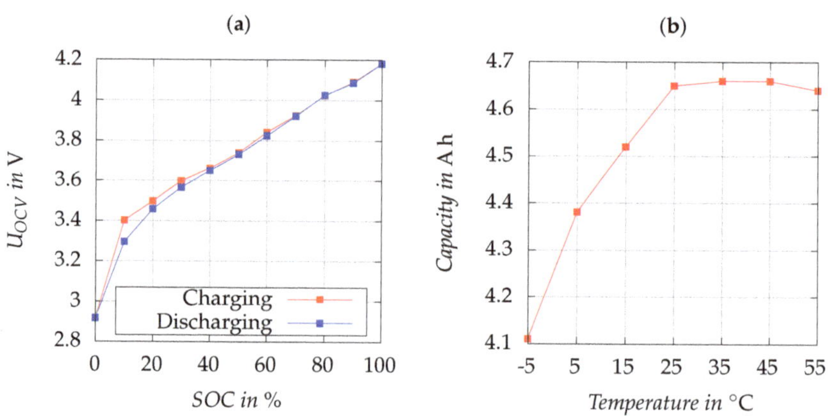

Figure 6. (**a**) Open circuit voltage U_{OCV} of the cell at 25 °C. (**b**) Change in cell capacity dependent to the temperature.

To validate our cell model and, subsequently, the SOC estimator, we established a comprehensive testing procedure. Initially, the test cell was fully discharged to its minimum voltage limit, establishing a reference point for the baseline SOC calculations. This was as follows: First, a complete charge–discharge cycle at a rate of C/3 to capture the cell's static characteristics. Subsequently, the cell was charged to various SOC levels (20%, 30%, 50%, 70%, 90%) and subjected to a dynamic current profile that was synthetically generated from empirical LEV riding data to emulate real-world conditions.

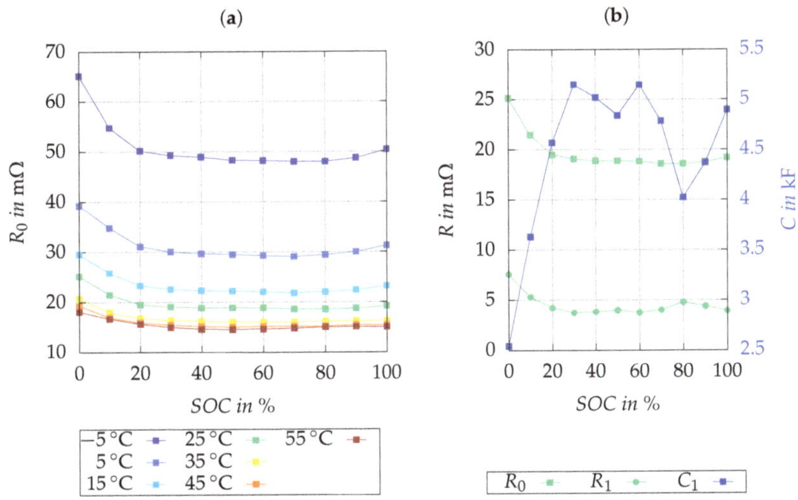

Figure 7. Results of the cell parametrization for R_0, R_1 and C_1. (**a**) Temperature dependency of R_0. (**b**) Change in R_0, R_1 and C_1 according to the SOC at 25 °C.

The validation process was paralleled by simulating the same test sequence using the developed cell model within the MATLAB/Simulink environment. The simulated cell voltages were then compared with the experimental data to evaluate the model's fidelity. In this study, we employed the root mean square error (RMSE) as the metric for quantifying the discrepancy between the simulated or estimated values $\hat{x}$ and measured or reference values x. Equation (9) describes the RMSE calculation, where n represents the count of data points.

$$\text{RMSE} = \sqrt{\frac{1}{n}\sum_{i=1}^{n}(x-\hat{x})^2} \tag{9}$$

Figure 8 visualizes the measured cell voltages U_{Cell} and the simulated cell voltages $\hat{U}_{Cell}$ for the validation test, as well as the corresponding simulation error. The RMSE of the voltage simulation is 7.09 mV. An exceptionally high error of 102.22 mV was observed outside of the operating range, notably when the cell was fully discharged. Despite this outlier, the simulation model's performance is robust, with the RMSE reflecting a high fidelity in capturing the cell's voltage dynamics. The model's performance remains consistent not only under constant current conditions, as noted during the initial charge/discharge cycle of the test, but also under dynamic load conditions. This robustness is observable in the second phase of the validation test, wherein the cell tested with realistic driving profiles of LEV across various levels of SOC.

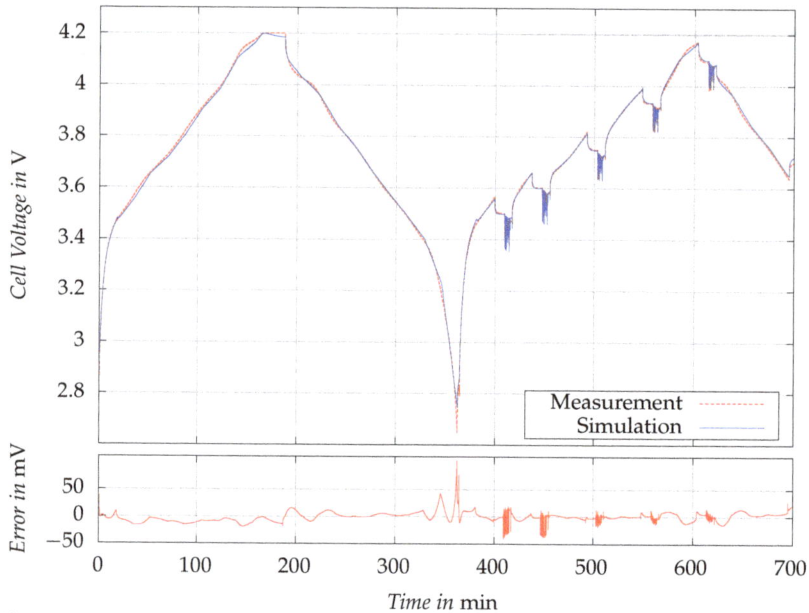

Figure 8. Battery model validation at 25 °C.

3.3. Results of EKF-Based State of Charge Estimator

Figure 9 presents the validation results of the developed EKF-based SOC estimator at the temperature of 25 °C. The graph compares the SOC values estimated by the EKF algorithm with the reference SOC values. The reference SOC values are derived using Coulomb counting method. The current is measured with a resolution of 1 mA and an accuracy of 0.1%. The test current profile employed here is identical to the one previously used to validate the cell model.

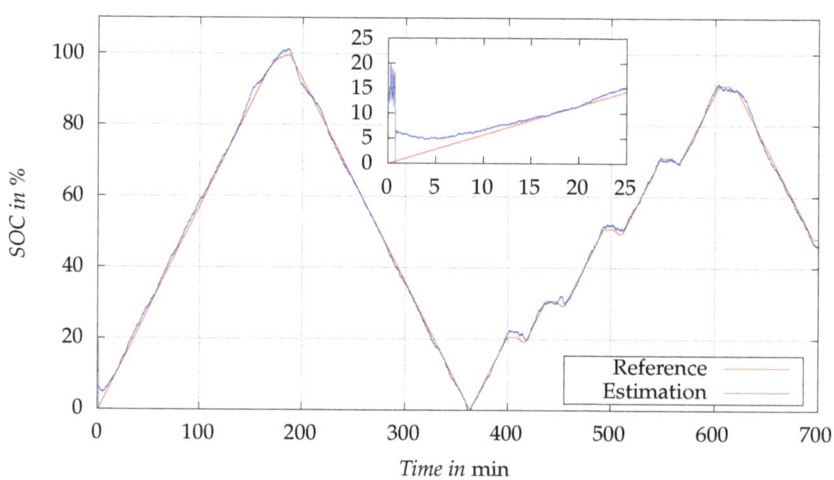

Figure 9. Validation of EKF-based SOC estimator at 25 °C.

Initially, the SOC estimation begins with an intentional discrepancy, starting at an estimated SOC of 20% when the actual reference SOC is at 0%. This initial error serves to test the robustness of the EKF algorithm in correcting errors caused by incorrect initial values. As the test progresses, the EKF-estimated SOC rapidly aligns with the reference SOC values, showcasing the estimator's capability to rectify the initial error.

In the zoomed inset, it is clearly evident that despite the substantial initial estimation error, the EKF algorithm quickly converges to the reference SOC. The close tracking between the estimated and reference values beyond this convergence point confirms the precision of the EKF algorithm in real-time SOC estimation under the specified test conditions. The RMSE of the test, including the initial value error, is 1.08%.

The algorithm's ability to accurately track the SOC, especially in the charging phase, enables precise control of the charging current, which is critical for the proposed SOC-dependent charging algorithm.

3.4. Aging Performance of the MCC Proposed Charging Algorithm

Figure 10 visualizes the aging test results, showing the change in SOH as a function of the number of cycles. The figure displays the minimum, maximum, and mean values for each measurement. We tested three cells under standard charging and two cells with MCC charging. The deviation between the cells in each group is negligible. The slight variation in SOH observed at the beginning of the tests between the cells subjected to standard charging, and those undergoing MCC charging can be attributed to the different calendar aging loads resulting from varying storage times of the cells. Notably, the standard charging test campaign started significantly earlier than the MCC charging test campaign. Furthermore, a slight drop in SOH of the cells tested with MCC charging after 850 cycles is observed, which results from an interruption in testing for eleven days due to maintenance work in the test field. During the maintenance work, the cells were disconnected and subsequently reconnected. The contacting is accompanied by certain contact resistances, which cause a slight scattering of measurement data of individual cells and presumably also contribute to the observed drop.

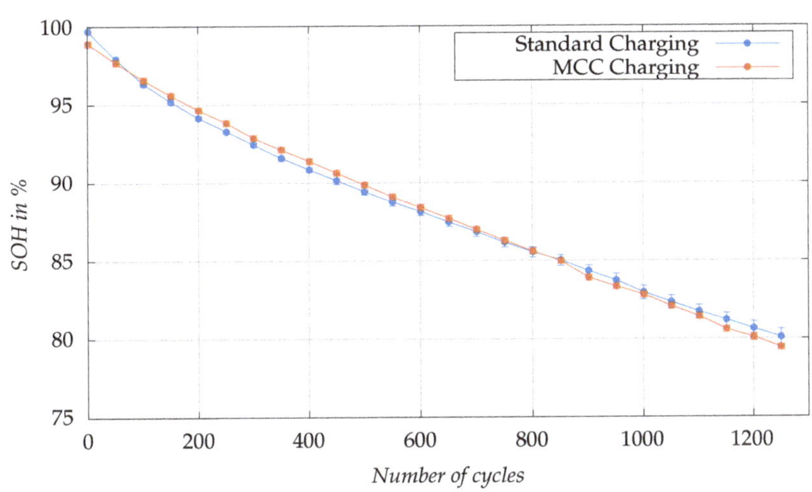

Figure 10. Aging test with standard and MCC charging at 25 °C.

The aging rate under MCC charging is slightly slower in the first part of the aging test (up to 90% SOH) compared to standard charging, and then becomes similar in the second part of the test. The end of life (EOL), determined as 80% SOH, was reached approximately 50 cycles sooner in cells subjected to MCC charging, when mean values are considered. Meanwhile, a cell subjected to standard charging reached EOL concurrently, at the 1200-cycle threshold. These observations support the expectation that the proposed MCC charging, despite its efficiency in reducing charging time, does not significantly affect the rate of aging compared to standard charging. The trends observed are consistent across the test for both charging methods, highlighting the successful transfer of insights from lab-scale experimental three-electrode cells to commercial cylindrical cells. This consistency demonstrates the scalability and practical applicability of the proposed MCC charging development method.

Figure 11 illustrates the change in total internal resistance ($R_0 + R_1$) at 50% SOC at 25 °C, throughout the aging process, for the proposed MCC charging protocol versus the standard charging protocol. In the mid-SOC range, the cell's resistance is more stable and exhibits a plateau. Therefore, 50% SOC is used for the resistance comparisons to reduce the impact of aging on SOC accuracy. The initial disparity in resistance and SOH values between the two tests, as previously discussed, is attributed to the varying storage time of the cells. Both sets of cells show a similar small decrease in internal resistance during the first cycles of the aging test. This is a typical effect of partial recovery after longer storage time. Then, both tests show an expected increase in resistance values over aging. The parallel trends observed indicate that the MCC charging algorithm does not induce additional Li-plating, which would be suggested by an unusual increase in resistance. These findings support our capacity test results, reinforcing that the MCC algorithm, designed for fast charging, preserves the integrity of the cell aging characteristics.

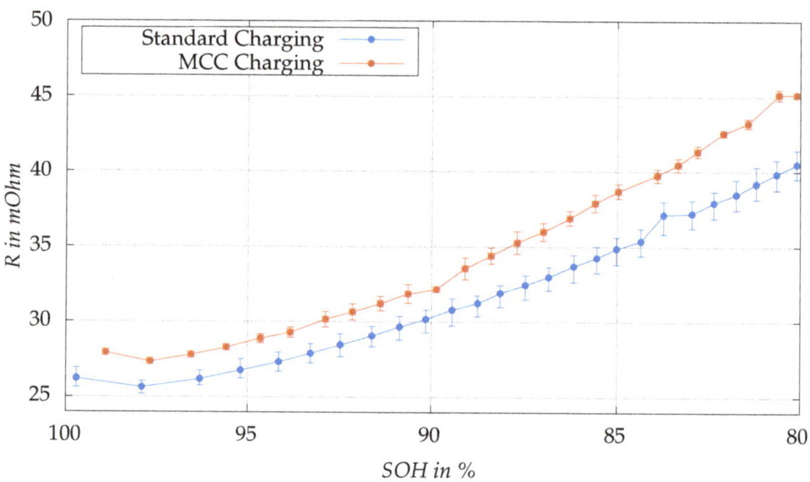

Figure 11. Changes in resistance (R_0 and R_1) at 50% SOC at 25 °C with aging under standard and MCC charging protocols.

4. Discussion and Conclusions

This paper significantly contributes to the field of battery MCC charging technology by enhancing the existing framework with the integration of a high-accuracy SOC estimator and by demonstrating its applicability to commercial cylindrical cells with NMC cell chemistry. The successful integration facilitates the transfer of precise SOC thresholds derived from three-electrode cell experiments to the commercial cell level, thereby enhancing practical applications and bridging the gap between experimental insights and industrial implementation.

We introduced an aging-optimized, SOC-dependent MCC charging algorithm designed to decrease charging times without accelerating battery degradation, by reducing the risk of Li-plating. This study underlines the importance of combining electrochemical analysis with modeling and estimation techniques to address the key challenges in battery charging. Our approach, leveraging SOC as a transfer parameter, ensures the scalability of laboratory findings to industrial applications.

The optimal charging pattern was determined using an experimental three-electrode cell, which allowed monitoring the anode potential to detect Li-plating. By incorporating the SOC thresholds derived from these experiments, the proposed MCC charging protocol offers enhanced stability compared to conventional voltage-based MCC protocols. This stability is rooted in the SOC-based approach is reduced susceptibility to factors that typically affect voltage-based protocols, such as temperature variations and electrochemical hysteresis.

To ensure proper control of the charging current in real-world applications, we developed an accurate State of Charge (SOC) estimator with a Root Mean Square Error (RMSE) of 1.08%, based on the Extended Kalman Filter (EKF), suitable for Battery Management Systems (BMS). Our experimental results demonstrate that our approach can decrease the time to reach 80% SOC by 30%, compared to the conventional constant CC-CV charging method, without accelerating the aging processes.

The integration of precise SOC thresholds from three-electrode experiments with an accurate SOC estimator facilitates the development of scalable charging guidelines for standard commercial cells. This method eliminates the need for physical reference electrodes during application and reduces the requirement for extensive cell testing in the charging protocol development phase.

In conclusion, the proposed aging-optimized, SOC-dependent MCC charging algorithm effectively reduces charging time while minimizing the risk of Li-plating, thereby preventing the acceleration of aging caused by fast charging. Although the study yields promising outcomes, incorporating temperature dependency into the MCC algorithm could potentially enhance charging efficiency and prolong battery life under variable environmental conditions. Future work will involve conducting three-electrode experiments at different temperatures to extend the charging profiles and incorporate temperature dependency. Consequently, subsequent research will explore the impact of temperature more thoroughly and investigate the underlying degradation mechanisms through post-mortem analyses, continuing to advance our understanding and application of efficient charging methodologies. This study contributes to the advancement of LIB charging technologies, particularly for the rapidly growing light electric vehicle (LEV) sector.

Author Contributions: This research was collaboratively carried out by a team of researchers, each contributing significantly to various aspects of the study. Conceptualization was jointly handled by A.K. and L.L. A.K. took the lead in creating the battery model and the SOC estimator, as well as in parameterizing and validating them. L.L. was responsible for the experimental investigation, which included building the three-electrode cells, designing, and conducting the tests, and extracting the SOC-dependent MCC Charging current. Both A.K. and L.L. managed the data curation process. The original draft of the manuscript was prepared by A.K. and L.L., with A.K. also leading the visualization efforts by generating graphs for the paper. The writing, review, and editing were supervised by C.K. C.K. and M.H. provided supervision for the entire project. All authors have read and agreed to the published version of the manuscript.

Funding: This research was funded by the Federal Ministry for Economic Affairs and Climate Action (BMWK) of Federal Republic of Germany as part of the Central Innovation Programme for SMEs (ZIM).

Data Availability Statement: Data are contained within the article.

Acknowledgments: This work was supported by the Helmholtz Association under the program "Energy System Design". The cell assembling and characterization was performed at KIT Battery Technology Center (KIT-BATEC). This work contributes to the research performed at the Center for Electrochemical Energy Storage Ulm & Karlsruhe (CELEST).

Conflicts of Interest: The authors declare no conflicts of interest. The funders had no role in the design of the study; in the collection, analyses, or interpretation of data; in the writing of the manuscript, or in the decision to publish the results.

References

1. Ziegler, M.S.; Trancik, J.E. Re-examining rates of lithium-ion battery technology improvement and cost decline. *Energy Environ. Sci.* **2021**, *14*, 1635–1651. [CrossRef]
2. Wu, H.; Alberts, G.; Hooper, J. *Battery Electric Vehicles: New Market, New entrants, New Challenges;* Technical Report; Deloitte LLP: London, UK, 2019.
3. IEA. Global EV Data Explorer: EV Stock, Cars, World, 2010–2020. 2021. Available online: https://www.iea.org/articles/global-ev-data-explorer (accessed on 28 Apirl 2022).
4. Korthauer, R. (Ed.) *Lithium-Ion Batteries: Basics and Applications;* Springer: Berlin/Heidelberg, Germany, 2018. [CrossRef]
5. Shen, W.; Vo, T.T.; Kapoor, A. Charging algorithms of lithium-ion batteries: An overview. In Proceedings of the 2012 7th IEEE Conference on Industrial Electronics and Applications (ICIEA), Singapore, 18–20 July 2012; pp. 1567–1572. [CrossRef]
6. Zhang, S.S. The effect of the charging protocol on the cycle life of a Li-ion battery. *J. Power Sources* **2006**, *161*, 1385–1391. [CrossRef]
7. Ghaeminezhad, N.; Monfared, M. Charging control strategies for lithium-ion battery packs: Review and recent developments. *IET Power Electron.* **2022**, *15*, 349–367. [CrossRef]
8. Lin, Q.; Wang, J.; Xiong, R.; Shen, W.; He, H. Towards a smarter battery management system: A critical review on optimal charging methods of lithium ion batteries. *Energy* **2019**, *183*, 220–234. [CrossRef]
9. Waldmann, T.; Kasper, M.; Wohlfahrt-Mehrens, M. Optimization of Charging Strategy by Prevention of Lithium Deposition on Anodes in high-energy Lithium-ion Batteries—Electrochemical Experiments. *Electrochim. Acta* **2015**, *178*, 525–532. [CrossRef]
10. Legrand, N.; Knosp, B.; Desprez, P.; Lapicque, F.; Raël, S. Physical characterization of the charging process of a Li-ion battery and prediction of Li plating by electrochemical modelling. *J. Power Sources* **2014**, *245*, 208–216. [CrossRef]
11. Williard, N.; He, W.; Hendricks, C.; Pecht, M. Lessons Learned from the 787 Dreamliner Issue on Lithium-Ion Battery Reliability. *Energies* **2013**, *6*, 4682–4695. [CrossRef]
12. Kaliaperumal, M.; Dharanendrakumar, M.S.; Prasanna, S.; Abhishek, K.V.; Chidambaram, R.K.; Adams, S.; Zaghib, K.; Reddy, M.V. Cause and Mitigation of Lithium-Ion Battery Failure—A Review. *Materials* **2021**, *14*, 5676. [CrossRef] [PubMed]
13. Vetter, J.; Novak, P.; Wagner, M.; Veit, C.; Möller, K.C.; Besenhard, J.; Winter, M.; Wohlfahrt-Mehrens, M.; Vogler, C.; Hammouche, A. Ageing mechanisms in lithium-ion batteries. *J. Power Sources* **2005**, *147*, 269–281. [CrossRef]
14. Hendricks, C.; Williard, N.; Mathew, S.; Pecht, M. A failure modes, mechanisms, and effects analysis (FMMEA) of lithium-ion batteries. *J. Power Sources* **2015**, *297*, 113–120. [CrossRef]
15. Schweidler, S.; de Biasi, L.; Schiele, A.; Hartmann, P.; Brezesinski, T.; Janek, J. Volume Changes of Graphite Anodes Revisited: A Combined Operando X-ray Diffraction and In Situ Pressure Analysis Study. *J. Phys. Chem. C* **2018**, *122*, 8829–8835. [CrossRef]
16. Waldmann, T.; Hogg, B.I.; Kasper, M.; Grolleau, S.; Couceiro, C.G.; Trad, K.; Matadi, B.P.; Wohlfahrt-Mehrens, M. Interplay of Operational Parameters on Lithium Deposition in Lithium-Ion Cells: Systematic Measurements with Reconstructed 3-Electrode Pouch Full Cells. *J. Electrochem. Soc.* **2016**, *163*, A1232–A1238. [CrossRef]
17. Li, Z.; Huang, J.; Yann Liaw, B.; Metzler, V.; Zhang, J. A review of lithium deposition in lithium-ion and lithium metal secondary batteries. *J. Power Sources* **2014**, *254*, 168–182. [CrossRef]
18. Liu, Y.H.; Hsieh, C.H.; Luo, Y.F. Search for an Optimal Five-Step Charging Pattern for Li-Ion Batteries Using Consecutive Orthogonal Arrays. *IEEE Trans. Energy Convers.* **2011**, *26*, 654–661. [CrossRef]
19. Dung, L.R.; Yen, J.H. ILP-based algorithm for Lithium-ion battery charging profile. In Proceedings of the 2010 IEEE International Symposium on Industrial Electronics, Bari, Italy, 4–7 July 2010; pp. 2286–2291. [CrossRef]
20. Luo, Y.F.; Liu, Y.H.; Wang, S.C. Search for an optimal multistage charging pattern for lithium-ion batteries using the Taguchi approach. In Proceedings of the TENCON 2009–2009 IEEE Region 10 Conference, Singapore, 23–26 November 2009; pp. 1–5. [CrossRef]
21. Liu, C.L.; Wang, S.C.; Liu, Y.H.; Tsai, M.C. An optimum fast charging pattern search for Li-ion batteries using particle swarm optimization. In Proceedings of the 6th International Conference on Soft Computing and Intelligent Systems, and The 13th International Symposium on Advanced Intelligence Systems, Kobe, Japan, 20–24 November 2012; pp. 727–732. [CrossRef]
22. Liu, Y.H.; Teng, J.H.; Lin, Y.C. Search for an optimal rapid charging pattern for lithium-ion batteries using ant colony system algorithm. *IEEE Trans. Ind. Electron.* **2005**, *52*, 1328–1336. [CrossRef]
23. Janakiraman, U.; Garrick, T.R.; Fortier, M.E. Review—Lithium Plating Detection Methods in Li-Ion Batteries. *J. Electrochem. Soc.* **2020**, *167*, 160552. [CrossRef]
24. Ansean, D.; Dubarry, M.; Devie, A.; Liaw, B.; Garcia, V.; Viera, J.; Gonzalez, M. Operando lithium plating quantification and early detection of a commercial $LiFePO_4$ cell cycled under dynamic driving schedule. *J. Power Sources* **2017**, *356*, 36–46. [CrossRef]
25. Petzl, M.; Danzer, M.A. Nondestructive detection, characterization, and quantification of lithium plating in commercial lithium-ion batteries. *J. Power Sources* **2014**, *254*, 80–87. [CrossRef]
26. Fan, J.; Tan, S. Studies on Charging Lithium-Ion Cells at Low Temperatures. *J. Electrochem. Soc.* **2006**, *153*, A1081. [CrossRef]
27. Petzl, M.; Kasper, M.; Danzer, M.A. Lithium plating in a commercial lithium-ion battery—A low-temperature aging study. *J. Power Sources* **2015**, *275*, 799–807. [CrossRef]
28. Sieg, J.; Bandlow, J.; Mitsch, T.; Dragicevic, D.; Materna, T.; Spier, B.; Witzenhausen, H.; Ecker, M.; Sauer, D.U. Fast charging of an electric vehicle lithium-ion battery at the limit of the lithium deposition process. *J. Power Sources* **2019**, *427*, 260–270. Erratum in *J. Power Sources* **2019**, *436*, 226846. [CrossRef]

29. Rangarajan, S.P.; Barsukov, Y.; Mukherjee, P.P. Anode potential controlled charging prevents lithium plating. *J. Mater. Chem. A* **2020**, *8*, 13077–13085. [CrossRef]
30. Epding, B.; Rumberg, B.; Mense, M.; Jahnke, H.; Kwade, A. Aging-Optimized Fast Charging of Lithium Ion Cells Based on Three-Electrode Cell Measurements. *Energy Technol.* **2020**, *8*, 2000457. [CrossRef]
31. Nölle, R.; Beltrop, K.; Holtstiege, F.; Kasnatscheew, J.; Placke, T.; Martin, W. A reality check and tutorial on electrochemical characterization of battery cell materials: How to choose the appropriate cell setup. *Mater. Today* **2020**, *32*, 131–146. [CrossRef]
32. Belt, J. *Battery Test Manual For Plug-In Hybrid Electric Vehicles*; U.S. Department of Energy: Washington, DC, USA, 2010. [CrossRef]
33. Plett, G.L. Extended Kalman filtering for battery management systems of LiPB-based HEV battery packs: Part 3. State and parameter estimation. *J. Power Sources* **2004**, *134*, 277–292. [CrossRef]
34. Rivera-Barrera, J.P.; Munoz-Galeano, N.; Sarmiento-Maldonado, H.O. SoC Estimation for Lithium-ion Batteries: Review and Future Challenges. *Electronics* **2017**, *6*, 102. [CrossRef]
35. Plett, G.L. *Battery Management Systems, Volume 2: Equivalent-Circuit Methods*; Artech House: Boston, MA, USA, 2016.
36. Xu, J.; Gao, M.; He, Z.; Han, Q.; Wang, X. State of Charge Estimation Online Based on EKF-Ah Method for Lithium-Ion Power Battery. In Proceedings of the 2009 2nd International Congress on Image and Signal Processing, Tianjin, China, 17–19 October 2009; pp. 1–5. [CrossRef]
37. Jiang, C.; Taylor, A.; Duan, C.; Bai, K. Extended Kalman Filter based battery state of charge(SOC) estimation for electric vehicles. In Proceedings of the 2013 IEEE Transportation Electrification Conference and Expo (ITEC), Dearborn/Detroit, MI, USA, 16–19 June 2013; pp. 1–5. [CrossRef]

batteries

MDPI

Article

A Physics–Guided Machine Learning Approach for Capacity Fading Mechanism Detection and Fading Rate Prediction Using Early Cycle Data

Jiwei Yao [1], Qiang Gao [2], Tao Gao [1], Benben Jiang [2,*] and Kody M. Powell [1,3,*]

[1] Chemical Engineering, University of Utah, Salt Lake City, UT 84112, USA; jiwei.yao@utah.edu (J.Y.); taogao@chemeng.utah.edu (T.G.)
[2] Department of Automation, Tsinghua University, Beijing 100084, China; qianggao@tsinghua.edu
[3] Mechanical Engineering, University of Utah, Salt Lake City, UT 84112, USA
* Correspondence: bbjiang@tsinghua.edu.cn (B.J.); kody.powell@chemeng.utah.edu (K.M.P.)

Abstract: Lithium–ion battery development necessitates predicting capacity fading using early cycle data to minimize testing time and costs. This study introduces a hybrid physics–guided data–driven approach to address this challenge by accurately determining the dominant fading mechanism and predicting the average capacity fading rate. Physics–guided features, derived from the electrochemical properties and behaviors within the battery, are extracted from the first five cycles to provide meaningful, interpretable, and predictive data. Unlike previous models that rely on a single regression approach, our method utilizes two separate regression models tailored to the identified dominant fading mechanisms. Our model achieves 95.6% accuracy in determining the dominant fading mechanism using data from the second cycle and a mean absolute percentage error of 17.09% in predicting lifetime capacity fade from the first five cycles. This represents a substantial improvement over state–of–the–art models, which have an error rate approximately three times higher. This study underscores the significance of physics–guided data characterization and the necessity of identifying the primary fading mechanism prior to predicting the capacity fading rate in lithium–ion batteries.

Keywords: Li–ion batteries; capacity prediction; feature extraction; data–driven; machine learning

1. Introduction

Lithium–ion batteries (LiBs) have become a key element in the transition to renewable energy due to their high specific energy and power density, falling costs, long life, low self–discharge rate, and lack of memory effect [1]. Many battery studies have focused on material synthesis, electrolyte development, and cell engineering. These efforts have advanced battery performance by controlling its design and manufacturing. A capacity degradation study is usually conducted to evaluate the performance of such a development. However, the degradation of LiBs is a lengthy process that poses a challenge to their development and deployment. Given the widespread use of LiBs, early prediction of their degradation behavior has become increasingly important. However, this task is challenging due to their nonlinear degradation over cycles, wide variability, and the necessity to meet specific operating conditions.

Model–based methods have been developed to predict the degradation behavior and address the early prediction problem. Model–based methods utilize models that describe batteries' internal chemical reactions and degradation mechanisms to predict the cycle life by considering the changes in specific mechanisms. These methods range from semi–empirical models to electrochemical models [2–4]. A challenge for these methods is parameterization, often addressed by parameter estimation algorithms such as the Kalman filter [5], and Particle filter [6]. Although these model–based methods have achieved some predictive success, developing comprehensive models for full cells to predict cycle life and determining model parameters for early cycle life prediction with limited data remains challenging due to the variety of degradation modes in a cell, which are coupled with mechanical and thermal heterogeneities [7].

Due to the complexity of utilizing model–based methods, statistical and machine learning methods have gained popularity for predicting battery lifetime from early cycle data thanks to recent advancements in algorithms, computational power, and data accessibility. Severson et al. developed a linear model–elastic net—that can accurately predict the battery cycle life with features extracted from the first 100 cycles [8]. By utilizing a more advanced regression model and undergoing sophisticated feature selection and engineering, Attia et al. achieved comparable or improved accuracy on the same dataset [9]. Gaussian process regression (GPR) has also become very popular for estimating battery degradation using early cycle data. Li et al. proposed a GPR model to predict the battery capacity and battery lifetime with features extracted from the dQ/dV curves from the first 30 cycles [10]. Richardson et al. proposed the integration of a conventional covariance–function–based GPR model with an explicit mean function, which can provide prior knowledge for lifetime prediction [11]. A support vector machine (SVM) has also been applied to estimate the lifetime [7,12,13]. Neural networks, with their good performance in solving nonlinear dynamic problems, have been applied widely in predicting degradation behavior. With features extracted and transformed from the capacity, voltage, current, and internal resistance recorded in the early cycle, Zhang et al. then employed a general regression neural network to predict the battery lifetime [14]. Xu et al. selected features from the discharge process and used a stacked denoising autoencoder to predict the battery lifetime [15]. Moreover, with a powerful deep neural network, given capacity, voltage, current, and temperature from the first cycle, the author was able to predict the battery life [16]. Yao et al. use autoencoder to extract features from multiple cycle follows with an elastic model to predict the battery lifetime given early cycle data [17]. However, neural networks, especially deep neural networks, are black box models and lack physics insight, leading to poor performance when extrapolating to regions not covered by the training data. Hybrid models, in which the core physics of the battery is combined with a data–driven model, have the potential to tackle such problems. Hybrid models are fast and have a low computational burden once trained by taking advantage of a physics–based model, including battery physics. Moreover, thanks to the core physics, these hybrid models' interpretability can help generalize the dataset and develop a physics–guided model [18]. Thelen et al. integrated experimental early degradation data with simulation data generated by a physics–based model to predict degradation behavior [19]. Instead of using the physics–based model to generate training data, by incorporating the physics constraint with the neural network's loss function, Najera–Flores et al. outperformed the model–based method and a machine learning method in battery remaining useful life prediction [20]. Furthermore, by combining a physics–based calendar and cycle aging model with a long short–term memory layer, Shi et al. proposed physics–informed machine learning for remaining useful life prediction [21]. Despite these pioneering works, in these previous studies, even though the cycle data are generated with varying operating conditions, which cover many degradation patterns, these studies attempted to propose a universal prediction model to perform a lifetime prediction, which may affect the model's predictability due to the underlying different fading mechanism. Some work has been conducted to address this issue. Severson et al. [8], Deng et al. [22], and Jiang et al. [23] proposed methods that can classify batteries into different groups with different capacity fading rates. However, these classification methods are data–driven and lack a physical explanation of the degradation patterns behind them. In contrast, recent works have correlated selected features and degradation modes with electrochemical explanations [24,25]. For example, Chen et al. demonstrated that the end–of–charge voltage is one of the strong indicators that distinguishes two major aging mechanisms: solid–electrolyte interphase (SEI) growth and Li plating [24]. With an understanding of different fading mechanisms, physics–guided electrochemical features can be extracted and create an explainable classification algorithm along with high accuracy. However, these studies lack predictive capability and fail to forecast the rate of degradation.

In this work, we propose a hybrid physics–guided and data–driven model to address a critical problem during battery operation: predicting capacity fading with early–stage data. Physics–guided features, based on a thorough understanding of battery physics, allow us to effectively identify degradation modes and accurately predict degradation behavior using early cycle data. These physics–guided features are integrated into our model to accurately reflect the physical and chemical states of the battery, thereby enhancing its predictive power and reliability. Moreover, with the understanding of battery physics, a physics–guided framework first determines the dominant fading mechanism of the battery through classification and then applies regression models tailored to each specific fading mechanism. Utilizing these features and the proposed framework, we develop a machine –learning approach that can predict the capacity fade rate with high accuracy. This model can significantly expedite the optimization of the battery charging protocol using early cycle data and reduce the test time for battery cycle tests, a common practice in battery development and deployment.

2. Data Generation

Commercial Li–ion coin cells (LIR2025H produced by EEMB Battery) with nickel manganese cobalt oxide (NMC) cathodes and graphite anodes are cycled on a Landt CT3001A battery test system to generate data for model development at different temperatures (Supplementary Materials Table S1). In each cycle, the battery is first charged by a constant–current constant–voltage (CCCV) method, i.e., the battery is charged at a pre–defined current until its voltage reaches a pre–defined cut–off voltage. Then, the charging voltage is fixed at this cut–off voltage until the charging current drops below C/50. After charging, the battery rests for 3600 s or 600 s and is then discharged to a cut–off voltage of 2.75 V at a constant current (CC), followed by a 600 s rest. Therefore, each cycle consists of four steps: CCCV charge, rest, CC discharge, and rest (Figure 1a). A total of 45 batteries are tested, and they are charged under different conditions during the CCCV charging with differing charging current and cut–off voltage but are discharged at the same rate (C/5) (Supplementary Materials Table S1). Two data streams from the raw cycle data are used for feature engineering and model development:

1. The first data stream contains the voltage of the battery, $V(t)$, during cycling (Figure 1a). The side reactions inside the battery, lithium deposits on the graphite anode, and the growth of the SEI cause the battery capacity to decay. Their effects are manifested in the shape of the $V(t)$ curve. Therefore, the features from the $V(t)$ and $I(t)$ curves contain physical information about side reactions. For example, incremental capacity analysis can be conducted, and features such as peak location and peak area can be used to identify the fading mode [26]. Note that $I(t)$ is predetermined as the input for battery cycling for constant current charge and discharge.
2. The second data stream contains the discharging capacity $C(N)$ and Faraday efficiency $\eta(N)$ for each cycle (Figure 1b). N refers to the battery cycle number. $C(N)$ is used to calculate the rate of capacity fading. The normalized capacity of each battery is shown in Figure 1b. Based on this data stream, the average capacity fading rate at the Nth cycle, ξ_N, is calculated as follows:

$$\xi_N = \frac{C_1 - C_N}{(N-1)C_1} \times 100\% \quad (1)$$

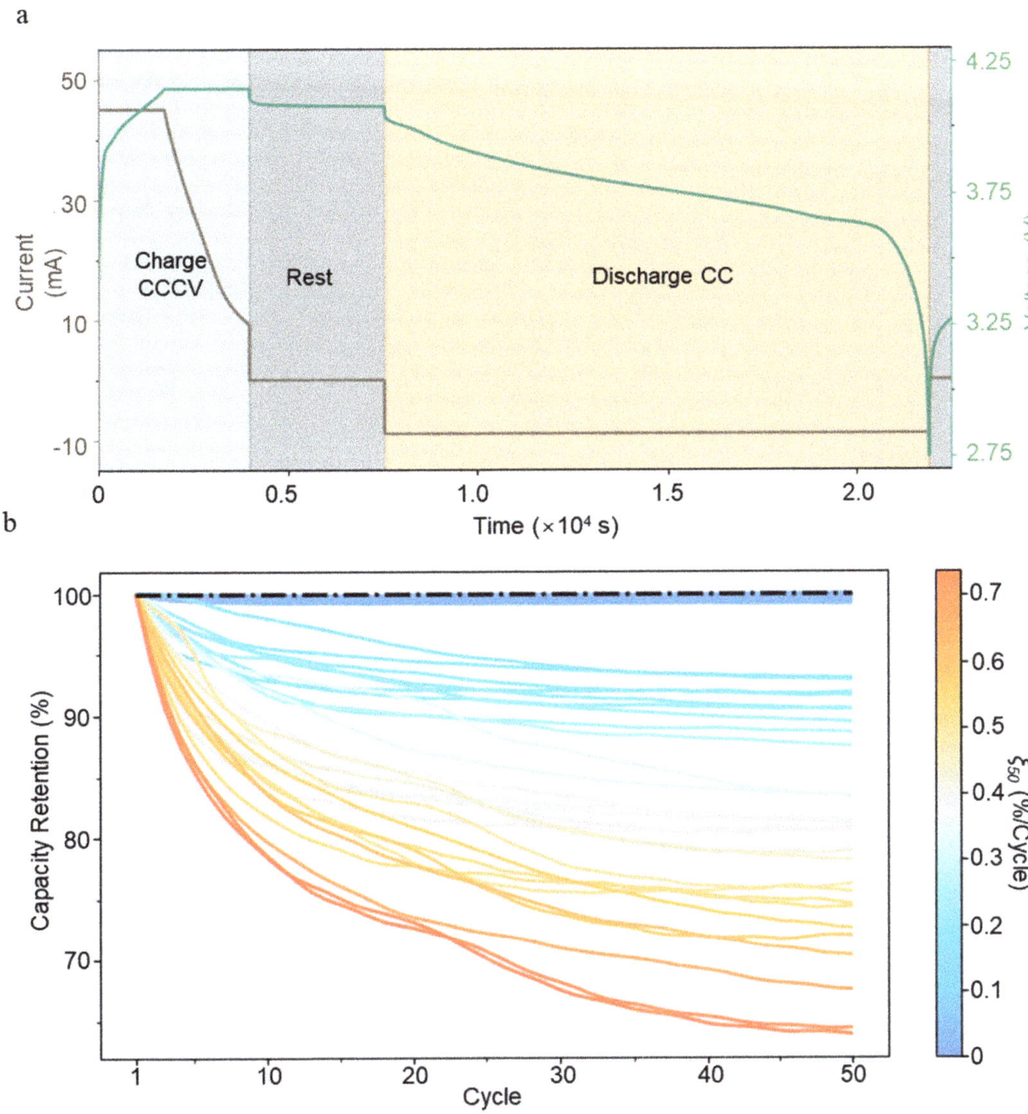

Figure 1. Various capacity fading curves under different cycling policies. Data streams show (**a**) the voltage and current profile of a cycle during battery cycling (Battery #40 at the 22nd cycle) and (**b**) normalized capacity retention during cycling for different batteries that are charged with different protocols. The color of each curve is scaled by the battery's average capacity fading rate at the 50th cycle.

3. Model Framework

The model consists of two modules: (1) the fading mechanism classification module, which classifies batteries charged under different protocols into two groups based on the dominant capacity fading mechanism, and (2) the average capacity fading rate prediction module, which separately predicts the average capacity fading rate of the batteries in each group (Figure 2). Li–ion batteries lose usable capacity during charge/discharge cycling, mainly due to the loss of active lithium ions. During charging/discharging, the active lithium ions carry charges and shuttle back and forth between the cathode/anode host, where side reactions occur. The loss of active lithium ions can be attributed to serval mechanisms: (1) SEI growth on graphite particles and (2) irreversible Li plating are the two most common reasons. SEI growth dominates under moderate conditions, and Li plating usually occurs at high charging rates. The current physical understanding does not provide an explicit boundary to distinguish the primary capacity fading mechanism for batteries cycled under different conditions. We show that such a boundary can be drawn with an unsupervised machine learning algorithm based on a feature that carries the physical information of the Li plating mechanism (fading mechanism classification module). With this algorithm, batteries will be classified into two groups: the Slow Fading Group, corresponding to batteries whose capacity fading is mainly governed by SEI growth, and the Fast Fading Group, corresponding to batteries whose capacity fading is dominated by Li plating. Two separate regression algorithms are built using features containing the physical information of the corresponding fading mechanism, which are used to predict the capacity fading rate for batteries in each group.

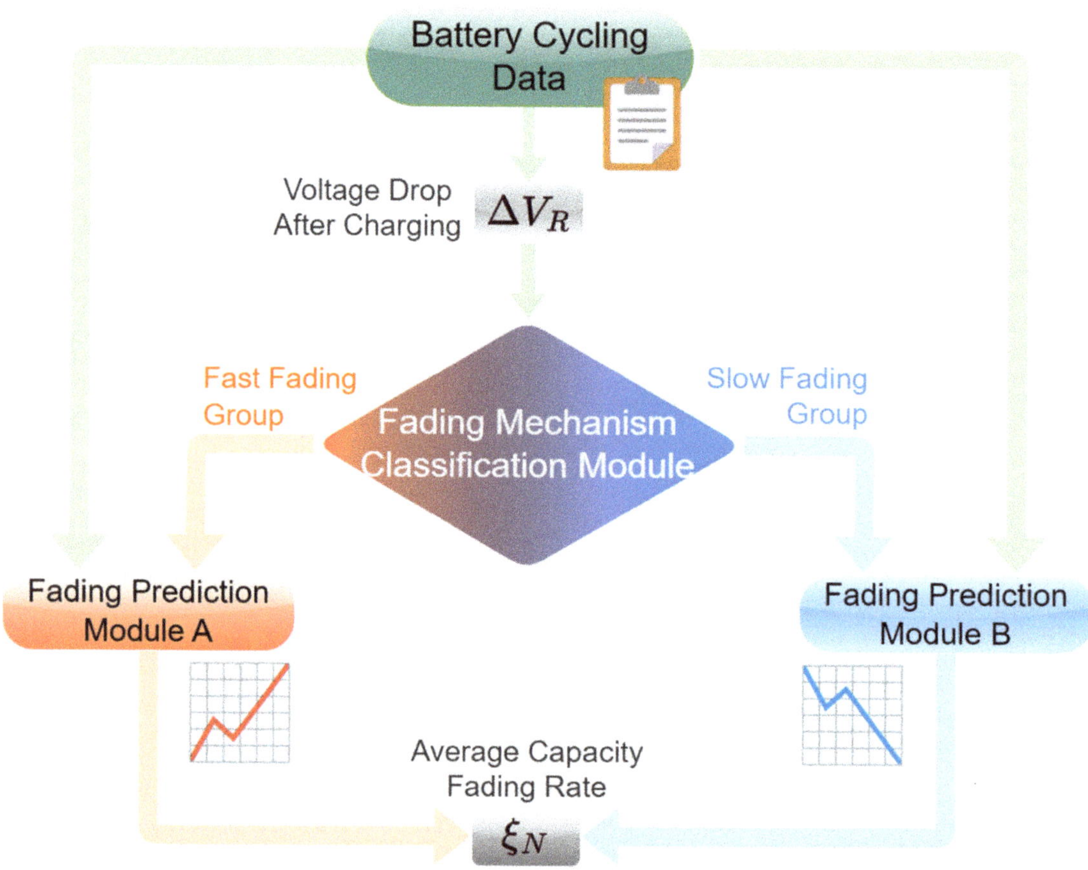

Figure 2. Data flow diagram of the proposed model. ΔV_R is extracted to classify the battery into the Fast Fading Group and Slow Fading Group. Then, relevant regression features are selected based on the classification results and are used to predict the average capacity fading rate. In this study, the average capacity fading rates at the 50th cycle are predicted.

4. Machine–Learning Model Development

4.1. K–Mean Clustering

Given an unlabeled dataset, the aim of K–means clustering is to partition the observation into k clusters, which minimizes the within–cluster sum of squares, such as variance. Given a set of datapoints, $X = \{x_1, x_2, x_3, \ldots, x_n\}$, K–means clustering aims to group the n datapoint into k sets $S = \{S_1, S_2, S_3, \ldots, S_k\}$. The objective can be expressed as

$$\arg\min_{S} \sum_{i=1}^{k} \sum_{x \in S_i} \|x - \mu_i\|^2 \tag{2}$$

where μ_i is the centroid of cluster i. The k value, which is the number of the cluster, will affect the K–means algorithm and needs to be specified before conducting any clustering analysis [27]. As mentioned before, depending on the operating conditions, the LiBs are subjected to different dominant degradation mechanisms. In this study, k is set to 2, inspired by the two dominant fading mechanisms: SEI growth and Li plating.

4.2. Linear Support Vector Regression

SVM is initially proposed by Cortes et al. to solve classification problems [28]. Drucker et al. adopted the principle of SVM and proposed a support vector regressor (SVR) [29]. SVMs have a solid theoretical foundation rooted in statistical learning theory and require only a small number of samples for training [30]. However, like any other model, the description and understanding of the data, including the number of dimensions, are crucial for model performance. Given a training dataset $D = \{(x_i, y_i), i = 1, 2, \ldots, n\}\left(x_i \in X = R^d, y_i \in Y = R\right)$, the linear SVR method maps the training dataset from the low–dimensional mapping space to the high–dimensional space through nonlinear mapping by solving the following primal problem:

$$\min_{w,b}\left(\frac{1}{2}w^T w + C\sum_{i=1}^{n} \max\left(0, \left|y_i - \left(w^T \phi(x_i) + b\right)\right| - \epsilon\right)\right) \tag{3}$$

where $w \in R^d$, ϵ is introduced to control sensitivity to loss which ignores errors less than ϵ, and C is a regularization term. The dual problem is formulated as

$$\begin{aligned} &\max_{a,a^*}\left(\sum_{i=1}^{n} (a_i - a_i^*)y_i - \sum_{i=1}^{n} (a_i + a_i^*)\epsilon - \frac{1}{2}\sum_{i=1}^{n}\sum_{j=1}^{n} (a_i - a_i^*)^T Q\left(a_j - a_j^*\right)\right) \\ &s.t.\begin{cases} \sum_{i=1}^{n} (a_i - a_i^*) = 0 \\ 0 \le a_i, a_i^* \le C, \forall i \end{cases} \end{aligned} \tag{4}$$

where $Q \in R^{n \times n}$ and the value within the matrix is calculated as $Q_{ij} = \phi(x_i)^T \phi(x_j)$. To evaluate the performance, the predicted average capacity fading rate should be compared to the actual fading rate. In this study, two performance metrics are considered: root mean square error (RMSE) and mean absolute percent error (MAPE), which are defined as

$$\text{RMSE} = \sqrt{\frac{1}{n}\sum_{i=1}^{n}(y_i - \hat{y}_i)^2} \tag{5}$$

$$\text{MAPE} = \frac{1}{n}\sum_{i=1}^{n} \frac{|y_i - \hat{y}_i|}{y_i} \times 100\% \tag{6}$$

5. Results

5.1. Fading Mechanism Classification

To develop a physics–guided machine learning algorithm for classifying batteries based on their primary capacity fading mechanism, the feature needs to contain the physical information of the fading mechanism. For Li–ion batteries, two mechanisms, SEI growth, and Li plating, contribute to the loss of active lithium ions and cause capacity fading. During charging, lithium ions are extracted from the oxide cathode, and meanwhile, lithium ions in the electrolyte near the anode are intercalated into the graphite anode. Under slow charging conditions (typically < C/3), the battery's capacity fading is mainly governed by the slow growth of SEI on the graphite anode, which gradually consumes the active lithium–ion inventory in the battery [31]. However, under fast charging conditions, a different capacity fading mechanism kicks in. When charged at a high rate, the insertion of Li into the graphite particles can outpace the diffusion of Li in graphite and lead to the saturation of the graphite particles on the surface [32]. As a result, Li plating occurs on the particles. During battery rest, a portion of the plated Li can still intercalate into the graphite, but some lose electrical contact with the graphite particles and become "dead". Therefore, Li plating consumes the Li inventory much faster than SEI growth, resulting in faster capacity fading.

Among these two data streams, $V(t)$ is a time series that reflects the battery's changing internal environment during cycling. More specifically, $V(t)$ is determined by the electrochemical potential of the electrode and the dynamics of various physical and chemical processes inside the battery. During the resting stage after charging, the change in voltage reflects the elimination of the Li insertion/extraction reaction at the interface, transportation of ions to balance the concentration inhomogeneity in the electrolyte, and diffusion of lithium ions to balance the chemical potential inhomogeneity in the electrode materials. When the battery enters the resting stage from the charging stage, the voltage profile usually consists of a sudden drop, which corresponds to the elimination of the reaction and ohmic overpotentials, followed by a gradual decay, which corresponds to the removal of the concentration or chemical potential gradient in the electrolyte and the electrode material (both oxide cathode and graphite anode). Given sufficient rest, the voltage reaches a plateau that corresponds to the equilibrium state of the battery. The voltage of the battery can be described by an electrochemical equation with F as Faraday's constant (Equation (7)) (see Supplementary Materials Note S2 for the derivation), in which the four terms correspond to the difference in surface chemical potential between the oxide cathode and the graphite anode, the concentration overpotential in the electrolyte, the ohmic overpotential in the electrolyte, and the reaction overpotential at the electrode/electrolyte interface. The last two terms vanish instantly once the battery enters the resting stage, and the second term tends toward zero as the electrolyte concentration gradient disappears. Without plated Li, the first term tends to reach the equilibrium voltage of the battery. In the graphite particles, the surface chemical potential ($\mu_d{}^g$) decreases because the diffusion of the inserted lithium ions toward the particle's interior reduces the lithium–ion concentration on the particle surface. However, if Li plates on graphite, the plated Li will react with graphite during the resting stage, so the surface lithium–ion concentration will remain constant until this reaction finishes. As a result, $\mu_d{}^g$ remains invariant, and the voltage drop only comes from the oxide, which will be reduced, compared to the case without Li plating.

$$V_{Bat} = -\frac{\mu_d^o - \mu_d^g}{F} + \frac{\mu_{Li^+}^o - \mu_{Li^+}^g}{F} + \left(\phi_l^o - \phi_l^g\right) + \left(\eta^o - \eta^g\right) \quad (7)$$

$$\Delta V_R = V_0 - V_{600} \quad (8)$$

Based on the above discussion, the voltage drops during the resting stage, ΔV_R, which is the voltage change within the first 600 s after CCCV charging, should effectively represent the capacity fading mechanism (Equation (8)). To confirm this, two batteries with drastically different capacity fade rates are chosen, and their $V(t)$ profiles during the resting stage are compared in Figure 3. The fast capacity fading of battery 33 is due to Li plating, whereas the slow capacity fading of battery 43 is due to SEI growth (Supplementary Materials Note S4). Evidently, their $V(t)$ profiles are distinct. Battery 43 (no Li plating) shows a smooth $V(t)$ curve with ΔV_R = 118.5 mV, whereas battery 33 (with Li plating) shows a zig–zag $V(t)$ profile with ΔV_R = 4.6 mV, which is two orders of magnitude lower. The zig–zag shape of the $V(t)$ profile is due to the limited number of data points being sampled in the experiment for a small ΔV_R. As discussed above, such a small voltage drop is because graphite particles with plated lithium fail to relax (reducing their surface lithium-ion concentration) during the resting stage, which occurs because the plated Li keeps intercalating with graphite particles and maintains its surface potential. A similar result has been observed in the experiment, where the voltage changes slowly during the rest period for graphite particles with plated lithium [32]. In contrast, for battery 43, without the intercalation of plated Li, the surface Li–ion concentration will gradually decrease, reducing the Li–ion concentration gradient within the graphite particle. Consequently, the additional relaxation in the graphite particle results in a much higher ΔV_R compared to the battery 33. A similar behavior can be observed when comparing their $V(t)$ profiles in other cycles (Supplementary Material Figure S3). Additionally, consistent differences are evident when comparing batteries with slow capacity fading (Supplementary Material Figure S5) to those with fast capacity fading (Supplementary Material Figure S4).

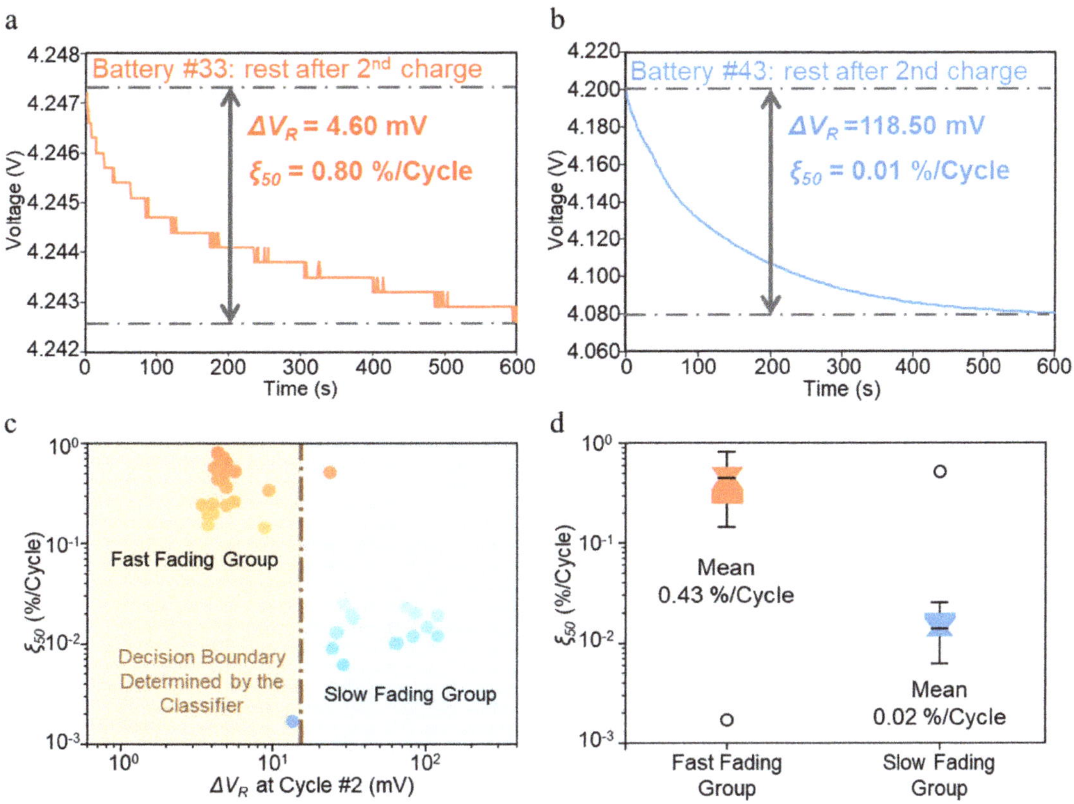

Figure 3. Samples of the voltage curve after the second charge and classification result with ΔV_R attained from the second cycle. Representative $V(t)$ profiles during the resting stage after the second

charge. The gradual decline in voltage after charging results from eliminating the concentration or chemical potential gradient in the electrolyte, cathode, and anode. In cells with Li plating, the plated lithium can intercalate into graphite, maintaining the Li–ion concentration at the graphite surface. This leads to minimal or no relaxation of the anode, resulting in a significantly smaller voltage drop compared to cells without Li plating. (**a**) A battery with Li plating is attained from battery 33 at the second cycle with a shallow voltage drop. (**b**) A battery without Li plating is attained from battery 43 at the second cycle with a large voltage drop. (**c**) Classification result with ΔV_R attained from the second cycle and corresponding decision boundary at 15.6 mV. (**d**) Distinctive distribution of ξ_{50} for two groups. Fast Fading Group has an average ξ_{50} at 0.43%/cycle, and the Slow Fading Group has an average ξ_{50} at 0.02%/cycle *. * Excluding two outliers.

The above examples suggest that ΔV_R could be a good feature for constructing a classification machine learning algorithm that can group batteries based on the capacity fading mechanism. Moreover, by measuring the voltage drop during the resting stage after constant voltage charging within a cycle, rather than the voltage value itself at a certain time, we can capture the process of relaxation and reduce the impact of different constant charging voltages. This approach allows us to eliminate the influence of different charging conditions and focus on ΔV_R to understand the cell's potential relaxation during the resting stage. To develop such an algorithm, ΔV_R of all the batteries at the 2nd cycle is computed and plotted on a log scale (Figure 3c). Visually, all data points fall into two groups: one located in the bottom–left region of the plot and another located in the top–right– area of the plot. Given a total of 45 cells, a boundary can be drawn by unsupervised machine learning using a K–means clustering algorithm (Figure 3c), which divides all batteries into two groups: Fast Fading Group (orange region) showing a $\Delta V_R < 15.6$ mV and Slow Fading Group (blue region) showing a $\Delta V_R > 15.6$ mV. To examine if this boundary can distinguish between batteries with different primary fading mechanisms, the capacity fade rate (ξ_{50}) of each group is plotted in a boxplot on a log scale (Figure 3d). Clearly, batteries classified into the Fast Fading Group (median = 0.45%/Cycle, mean = 0.43%/Cycle) have significantly higher capacity fading rates than batteries in the Slow Fading Group (median = 0.01%/Cycle, mean = 0.02%/Cycle). By treating outliers as incorrectly classified, given the second cycle data, the classification achieves a 4.3% error rate. A similar classification can be achieved using only the $V(t)$ data from the other early cycles (Supplementary Materials Figure S6), indicating the reliability of the ΔV_R feature. As the batteries cycle, for batteries with Li plating as the dominant fading mechanism, the usable capacity decreases quickly, resulting in a higher real C–rate, which makes the lithium plating servers. Therefore, the classification result improves with the ΔV_R collected from a later cycle (Supplementary Materials Figure S6). These results demonstrate that using ΔV_R as a single feature, an unsupervised machine learning module can accurately cluster batteries into two groups: a Fast Fading Group with Li plating as the main fading mechanism and a Slow Fading Group with SEI growth as the main fading mechanism. Being able to identify the main capacity fading mechanism is a critical step in realizing the accurate prediction of battery capacity during cycling for two reasons: (1) it enables feature selection based on the fading mechanism without having to resort to the time–consuming feature engineering process in a typical machine learning model development; (2) by carefully examining the fading mechanism, features containing the most relevant information can be identified, which can increase the prediction accuracy of the model. In the next section, we discuss how we choose such physics–guided features to construct a machine learning algorithm to predict battery capacity fading based on the dominant fading mechanism.

5.2. Capacity Fading Prediction

With the unsupervised Fading Mechanism Classification algorithm, the batteries are clustered into two groups using a single feature from the second cycle. For batteries in the Fast Fading Group, their capacity loss is mainly caused by the "dead" Li, which forms during charging under a high charging rate. Although ΔV_R is determined by the evolution of the surface lithium–ion concentration in both the oxide cathode and graphite anode, the contribution from the graphite anode is negligible because its surface concentration remains invariant due to the reaction between the plated Li and graphite. Therefore, a much smaller ΔV_R is observed than in batteries without Li plating. With an invariant electrochemical potential from the graphite electrode, the voltage drops after charging comes from the relaxation of the oxide. Additionally, a larger ΔV_R indicates a steeper concentration gradient in the oxide cathode after charging, corresponding to a more stringent charging condition and more severe Li plating. Therefore, it is expected that ΔV_R correlates positively with the average capacity fading rate. This conclusion, drawn from the physical analysis, is supported by the correlation analysis shown in Figure 4a. Figure 4a shows the correlation between the average capacity fading rate and ΔV_R for all batteries in the Fast Fading Group. For most of the region, the correlation is greater than 0.75, which indicates that a battery with a larger ΔV_R will have a larger ξ_N at the Nth cycle. This positive correlation corresponds to our physics insight that a larger ΔV_R is related to more plated Li and a larger ξ_N. However, in some regions, particularly with ΔV_R obtained from the first four cycles, the correlation is relatively low due to the battery experiencing a relatively weak Li plating reaction. In the early cycles, Li plating does occur in these batteries; however, other fading mechanisms also contribute to degradation, affecting the correlation between ΔV_R and the capacity fading rate. As degradation progresses and Li plating becomes more dominant, the correlation strengthens (see Supplementary Materials Note S8). As the actual capacity decreases, the actual charging environment will deteriorate for a battery dominated by Li plating, given a fixed charging current and an increased actual C–rate. In other words, the relationship between the average capacity fading rate and ΔV_R only develops after the situation worsens sufficiently. In this case, it occurs after the first four cycles. Overall, a mean correlation of 0.76 is observed after excluding the first four cycles, indicating a strong positive correlation between ΔV_R and ξ_N. Therefore, ΔV_R is used as a key feature in constructing a regression model (Fading Prediction Module A) to accurately predict the capacity fading rate for batteries in the Fast Fading Group.

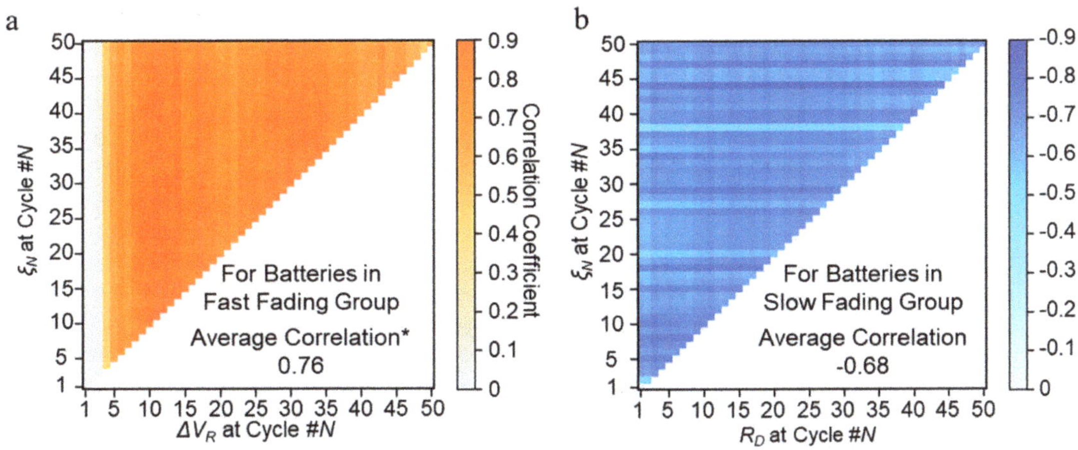

Figure 4. High correlation between ξ and selected features (ΔV_R and R_D). (**a**) Correlation plot between ξ_N and ΔV_R from cycle #1 to 50 with an average of 0.76, excluding the first four cycles. (**b**) Correlation plot between ξ_N and R_D from cycle #1 to 50, with an average of –0.68. * Excluding the first four cycles.

For batteries in the Slow Fading Group, given that their capacity fading is governed by SEI growth, the best feature for predicting capacity fading should contain the characteristics of the SEI growth physics. During battery cycling, the SEI grows by irreversible decomposition of the electrolyte. This process consumes the usable lithium ion in the battery and leads to capacity loss. Meanwhile, as the SEI grows thicker, its resistance to current, R_{SEI}, increases accordingly. Therefore, more capacity loss means a higher R_{SEI}. In other words, R_{SEI} correlates positively with the capacity fading rate. Since R_{SEI} cannot be measured directly during battery cycling, an equivalent resistance must be computed from the $V(t)$ profiles to approximate R_{SEI}. For this purpose, we first examine the physics governing the battery voltage (Equation (9)). During discharge, the battery voltage is the difference between the equilibrium voltage V_{Eq} and the total overpotential, including the ohmic, the reaction overpotential, the electrolyte concentration overpotential, and the SEI overpotential. The former three can be assumed to be invariant between cycles, given that SEI growth has a negligible influence on them. In addition, V_{Eq} is also invariant between cycles since all the batteries are discharged from the same SOC. Therefore, the $V(t)$ profile change during the discharge stage is only attributed to SEI growth. For this reason, the computed average resistance during the discharge process is used to approximate the R_{SEI} (Equation (10)). With these two equations, we can derive and conclude that there is a linear negative relationship between R_D and R_{SEI} (Equation (11)).

$$V_{Bat} = V_{Eq} - \eta_{ohim,e} - \eta_{rxn} - \eta_{e} - \eta_{SEI} \tag{9}$$

$$R_D(N) = \frac{1}{t}\int_0^t \frac{V_{Bat}(N)}{I_D}\mathrm{d}t \tag{10}$$

$$R_D(N) = \frac{1}{t}\int_0^t \frac{C - \eta_{SEI}(N)}{I_D}\mathrm{d}t = C - \frac{\eta_{SEI}(N)}{I_D} \sim -R_{SEI}(N) \tag{11}$$

Based on the above physical analysis, R_D is the average resistance during the discharge process. For batteries in the Slow Fading Group, the polarization effect from charging diminishes much faster compared to the Fast Fading Group. Consequently, after the resting stage, the polarization effect from charging can be disregarded. However, the polarization progressively increases during discharge. To minimize the impact of discharge–induced polarization, R_D is calculated using the data from the first three seconds. It is expected that R_D correlates linearly with ξ_N. To verify this assumption, for all batteries in the Slow Fading Group, a correlation plot between R_D and ξ_N at different cycles is presented (Figure 4b). For most regions, the correlation between R_{SEI} and ξ_N is lower than –0.68, confirming the strong negative correlation between R_D and capacity fading. For this reason, R_D will be used as the feature to construct a regression model (Fading Prediction Module B) to predict ξ_N for batteries classified as the Slow Fading Group.

Next, we present the machine learning modules with ΔV_R and R_D as the features to predict the capacity fading for batteries in either the Fast Fading Group or Slow Fading Group and show their prediction results. The objective of the regression model is to forecast the average capacity fading rate at the 50th cycle (ξ_{50}) of a battery charged with a specific charging protocol using data from early cycles. The model uses ΔV_R (for batteries in the Fast Fading Group) and R_D (for batteries in the Slow Fading Group). Due to the previously mentioned poor relationship between ξ_{50} and ΔV_R for the Fast Fading Group in the first four cycles, R_D and ΔV_R are obtained from the fifth cycle.

Such an assessment of the charging protocol using data only from the early cycles can significantly improve the efficiency of the optimization of the charging protocol and shorten the cycle test time. Given the limited dataset, linear SVR is used. SVR can balance the model complexity and prediction error by incorporating an ε–insensitive region within the optimization problem, providing excellent generalization ability and high prediction accuracy. In comparison, a deep learning model, such as a deep neural network involving many training parameters, needs a large dataset for model training, which is not applicable in this study. In this study, a 70/30 training/test split on the dataset was used. The training data contains 19 batteries from the Fast Fading Group and 10 batteries from the Slow Fading Group. The testing data includes nine batteries and five batteries from the Fast Fading

Group and Slow Fading Group, respectively. Furthermore, to avoid over–fitting on the training dataset, k–fold cross–validation was performed to choose the hyper–parameters. Various k values are evaluated, with the results showing that 5–fold cross–validation provides the best result.

To demonstrate the predictability of the physics–guided features, single—feature regression models are developed with ΔV_R and R_D from the fifth cycle, respectively. A previously reported model, the Variance model, which also utilizes one feature, is used as the benchmark to validate the proposed framework. According to Severson et al., the log variance of the discharging curve difference $\Delta Q_{N\text{-}1}(V)$ is selected due to its strong correlation with the predicted features [8]. The parity plots of the actual and predicted ξ_{50} are demonstrated in Figure 5a (Fading Prediction Module A for the Fast Fading Group and Fading Prediction Module B for the Slow Fading Group), while Figure 5b presents the error distribution. In contrast, for the benchmark model, the parity plot and the prediction error are presented in Figure 5c and Figure 5d, respectively. According to parity plots, the Variance model performs relatively poorly because of the high prediction error for batteries with a low capacity fade rate. This indicates that the selected features, log $\Delta Q_{5\text{-}1}(V)$, cannot determine the dominant fading mechanism. This error is also shown by the distribution of the mean absolute percentage error (MAPE) in the histogram in Figure 5d. In the Variance model, for batteries classified as slow fading in the test set, the maximum absolute percentage error reaches 1201.13%, and the corresponding mean MAPE is 480.69%. The poor performance indicates that the benchmark fails to make capacity fading rate predictions for batteries classified in the Slow Fading Group. In contrast, the proposed model can avoid this problem by first discerning the dominant capacity fading mechanism. By selecting specific physics–guided features, ΔV_R and R_D, for each group, we achieve an overall test MAPE of 27.96%. In contrast, because the selected features, log $\Delta Q_{5\text{-}1}(V)$, cannot determine the dominant fading mechanism, the Variance model has a 192.61% test MAPE, which is significantly higher than that of the proposed framework.

In addition to the selected features, other physics–guided features have been added to improve prediction accuracy further. For LiBs, the Faraday efficiency describes the efficiency with which electrons are transferred in batteries. Therefore, it can be used to tell how many electrons are consumed by the side reactions which are directly related to capacity loss. By assuming a linear relationship between the geometric average Faraday efficiency of cycle N_1 and N_2, we can calculate the prospective cumulative Faraday efficiency of cycle N_2 to reflect capacity loss (see Supplementary Materials Note S3). Moreover, the average capacity fading rate at the fifth cycle, ξ_5, is added as the third feature to provide the model with previous fading information.

Two state–of–the–art models are introduced [8,14] to validate the proposed method. These two models target early–stage battery lifetime prediction. In contrast to the proposed model, these models do not have a dominant fading mechanism classifier. Therefore, relevant features are extracted following prediction. However, at the same time, the selected features and proposed models are different (see Supplementary Material Table S3).

(1) Discharge model. In [8], the authors proposed three models to predict battery lifetime. The model with six features, derived from first–100–cycle data, provided the best prediction result with an elastic net regression model. We reproduce these features using the first–5–cycle degradation data.
(2) General Regression Neural Network (GRNN). In [14], based on the importance ranking of all features, four features derived from first–100–cycle data were selected and input into a GRNN model. However, the fourth feature, the increase in internal resistance, is not included in our dataset. Therefore, three features are reproduced with our dataset and fed into a GRNN model.

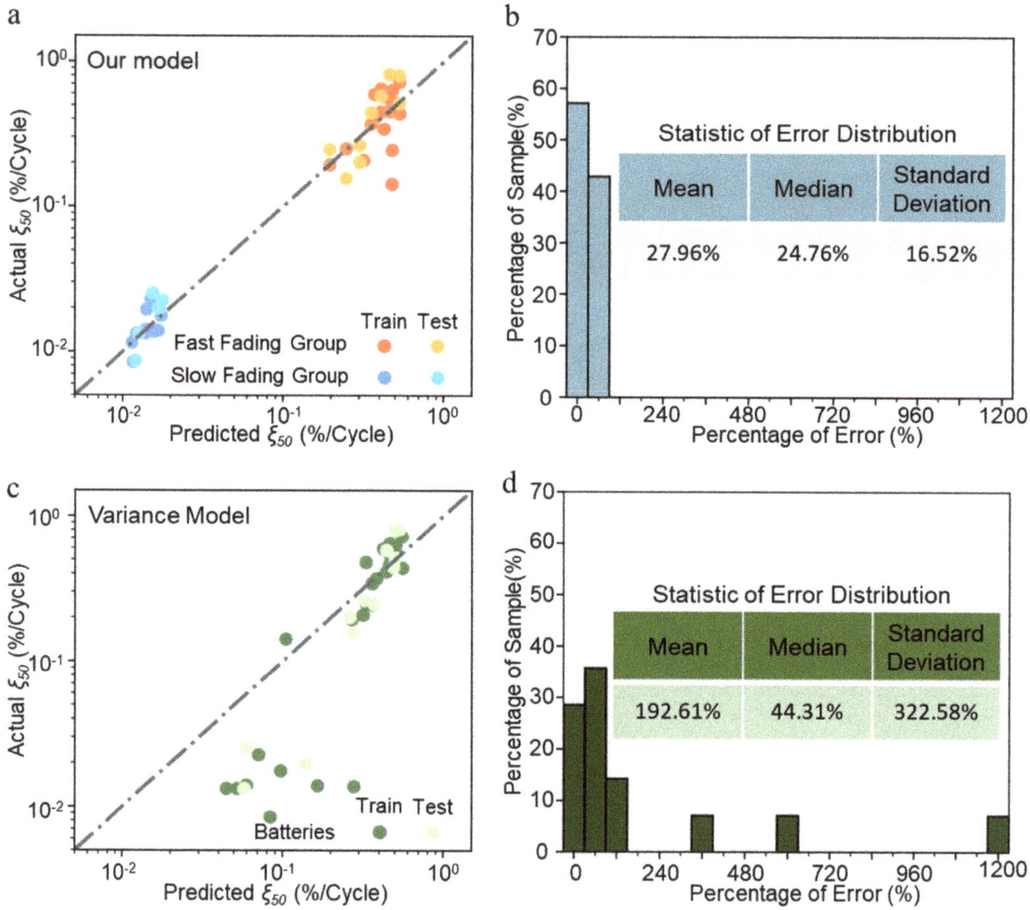

Figure 5. Prediction comparison between the proposed model and the Variance model (benchmark). (**a**) Parity plot of predicted ξ_{50} vs. actual ξ_{50} for SVR regression of Fading Prediction Module A (Fast Fading Group) using only ΔV_R obtained from the fifth cycle, Fading Prediction Module B (Slow Fading Group) with R_D obtained from the fifth cycle. (**b**) Distribution of the test MAPE of the proposed model which has a mean of 27.96%, median of 24.76%, and standard deviation of 16.52%. (**c**) Parity plot of predicted ξ_{50} vs. actual ξ_{50} for SVR regression of the Variance model with log variance of $\Delta Q_{5-1}(V)$. It fails to predict batteries with a low capacity fading rate. (**d**) The distribution of the test MAPE of the Variance model with a mean of 192.61%, median of 44.31%, and standard deviation of 322.58%. The prediction performance of the variance model is poorer than that of the proposed model according to the distribution of the test MAPE.

Tree–based regression has also been used to forecast battery lifespan [33]. The proposed framework, which includes a clustering method to determine a primary fading mechanism followed by regression, is similar to tree–based regression. Consequently, a random forest regression (RFR) technique has been incorporated as a benchmark with ΔV_R, R_D, prospective cumulative Faraday efficiency, and ξ_5. The prospective cumulative Faraday efficiency and ξ_5 are added to the Variance model, which denotes as Variance–M. Furthermore, three naïve models are also added (Supplementary Materials Table S2). To ensure a robust model evaluation, we performed 10 random train–test splits. A comparison of the mean test errors between the proposed framework and other models is summarized in Figure 6. With the help of two extra features, prospective Faraday efficiency and ξ_5, for the proposed method, the test MAPE is reduced to 17.09% while the test RMSE is at 0.09%/cycle. In contrast, the Discharge model only achieves a test MAPE of 206.15% and a test RMSE of 0.11%/cycle. The GRNN model has a test MAPE of 73.13% and a test RMSE of 0.11%/cycle. The RFR performs relatively well, which suggests that it is important to separate the battery. Among the nine models, the proposed method has the lowest test MAPE and outperforms all other benchmarks, including three naïve models. As for the test RMSE, the Variance–M provides the lowest RMSE of 0.08%/cycle while the proposed model has a comparable value at 0.09%/cycle. The relatively low performance at the test RMSE can be attributed to the distinctive and large difference in fading behavior between the Fast Fading Group and the Slow Fading Group. The RMSE is skewed by the high average capacity fading rate in the Fast Fading Group. Therefore, the Variance–M model tends to fit the batteries with a higher average fading rate to lower the fitting error while accepting a significant percentage error in the Slowing Fading Group, resulting in a small test RMSE and large test MAPE (see Supplementary Materials Table S3).

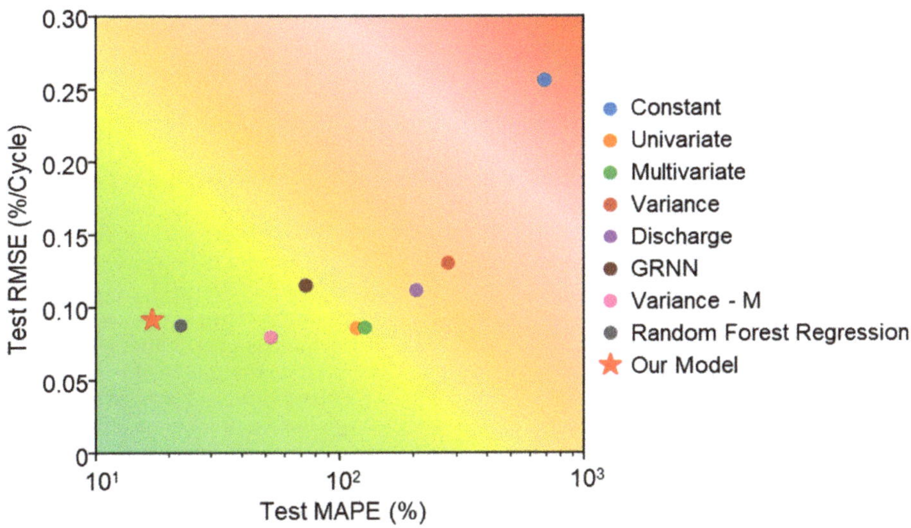

Figure 6. The proposed model outperforms the benchmarks (Variance Model [8], Discharge Model [8], and GRNN [14]). The proposed method provides the lowest test MAPE of 17.09%, and the corresponding test RMSE is 0.09%/cycle. The Variance–M model achieves the lowest test RMSE of 0.08%/cycle and comes with a test MAPE of 52.69%.

6. Discussion

Our work aims to use the physical mechanism to guide the feature selection and use a data–driven method to describe the relationship between physics–based inputs, (1) ΔV_R and the dominant capacity fading mechanism and (2) ΔV_R and R_D, and the target output, ξ_N. In our work, inspired by the two different fading mechanisms within Li–ion batteries, a fading mechanism classification is performed to determine which fading mechanism is dominant before making a capacity fading prediction. However, it is worth noting that there are cases where the classifier fails to describe the dominant fading mechanism. As mentioned in the previous paragraph and shown by many studies, for Li plating to occur, the battery must operate at a high current ratio with respect to its present capacity and active lithium–ion concentration [32,34–37]. For a fresh battery, due to its relatively small loss of active Li–ion, the operating environment is not harsh enough to reduce and deposit sufficient Li, which leads to an incorrect classification result. However, this situation only occurs in a few batteries and can be avoided by selecting ΔV_R from later cycles. The selected feature ΔV_R provides a distinctive classification result even with data obtained from the second cycle and achieves 95.6% accuracy. This highlights the importance of physics–guided insights in model development. Due to the different characteristics of these two fading mechanisms, growth of SEI and Li plating, physics–guided data characterization is performed, and two features, ΔV_R and R_D, are extracted from the cycle data. Along with other physical features, prospective cumulative Faraday efficiency, and past average capacity fading rate, regression models for each group have a good performance given five cycles data, compared to that of the Variance–M, which uses the log variance of $\Delta Q_{N-1}(V)$ for all batteries. The reason for the accurate results and good extrapolation of these regression models with selected features can be summarized by two factors. (1) Thanks to the dominant fading mechanism classifier, in the proposed model, each regression module is responsible for one fading mechanism with corresponding physics–guided features. In contrast, the Variance–M tried to generate a model that fit both fading mechanisms, leading to poor performance for batteries classified as the Slow Fading Group. (2) The physics–guided data characterization provides the model with good extrapolation and generalizability. Therefore, even though only five cycles are given, the regression model still outperforms all benchmarks. Moreover, the selected features can be easily obtained in the filed application, ΔV_R can be obtained from the short rest immediately after the high constant current charging, while R_D can be acquired once the battery is discharged after a rest period.

7. Conclusions

In this study, we proposed a physics–guided, data–driven model designed to predict capacity fading under fast charge conditions for Li–ion batteries. With an understanding of battery physics, the framework includes a classifier to identify the dominant battery fading mechanism and two regressors tailored to predict battery capacity fading based on the identified mechanism. Inspired by the two capacity fading mechanisms, SEI growth, and Li plating, and guided by an understanding of battery physics, we use a physics–guided feature, ΔV_R, which represents the voltage drop during the rest period after charging. This feature effectively captures the potential relaxation process and is used to identify the dominant fading mechanism. We then classify the batteries into two groups: the Fast Fading Group and the Slow Fading Group. With the ΔV_R obtained from the second cycle, we achieve an accuracy of 95.6%. Following the clustering, there is a corresponding capacity fading prediction model for each fading mechanism. With two additional physics–guided features, the proposed method achieved an overall test MAPE of 17.09% and test RMSE of 0.09%/cycle. In contrast, the state–of–art models, the Discharge model, and GRNN, have a test MAPE of 206.15% and 73.13%, respectively. The poor performance of these two models is caused by the inability to discern the dominant fading mechanism, which leads to poor performance of batteries with a slow fading rate. Moreover, because the model is based on electrochemical principles, it is expected that physics–guided features can still capture fading mechanisms such as SEI growth and Li plating. Consequently, the proposed model has the potential to be applied to other types of lithium–ion batteries with different chemistries, which remains a subject for future validation. Moreover, other fading modes, such as the loss of active materials and conductivity loss, should also be considered in future work.

Supplementary Materials: The following supporting information can be downloaded at: https://www.mdpi.com/article/10.3390/batteries10080283/s1, Note S1: Battery Activation, Figure S1: Example of determination of activation, Note S2: Battery Voltage Derivation, Note S3: Feature Formulation, Note S4 and Figure S2: Dissemble of Batteries Clustered as Slow Fading Group and Fast Fading Group, Note S5 and Figure S3: ΔV_R at Different Cycles, Note S6, Figure S4, and Figure S5: ΔV_R for Different Batteries at Same Cycles, Note S7 and Figure S6: K-mean Clustering results with Data Obtained From Early Cycle (1-6), Note S8 and Figure S7: Evolution of ΔV_R, Table S1: Experiment Conditions, Table S2: Feature Selection, Table S3: Model Performance.

Author Contributions: Conceptualization, J.Y., Q.G., T.G. and B.J.; methodology, J.Y., Q.G., T.G. and B.J.; software, J.Y. and Q.G.; validation, J.Y. and T.G.; formal analysis, J.Y., Q.G. and T.G.; investigation, J.Y., Q.G., T.G., B.J. and K.M.P.; resources, J.Y., Q.G., T.G., B.J. and K.M.P.; data curation, J.Y., Q.G. and T.G.; writing—original draft preparation, J.Y., Q.G. and T.G.; writing—review and editing, J.Y., T.G., B.J. and K.M.P.; visualization, J.Y. and T.G.; supervision, T.G., B.J. and K.M.P.; project administration, T.G., B.J. and K.M.P.; funding acquisition, T.G., B.J. and K.M.P. All authors have read and agreed to the published version of the manuscript.

Funding: National Natural Science Foundation of China: 62273197. Tsinghua-Toyota Joint Research Fund.

Data Availability Statement: The battery datasets used in this study are available at https://data.mendeley.com/datasets/m8w8sjk3vm.

Acknowledgments: The authors would like to acknowledge the support of Ruixue Liu for the revision. B. Jiang was supported by the National Natural Science Foundation of China under Grant 62273197 and the Tsinghua–Toyota Joint Research Fund.

Conflicts of Interest: The authors declare no conflicts of interest.

References

1. Xiong, W.; Mo, Y.; Yan, C. Lithium–Ion Battery Parameters and State of Charge Joint Estimation Using Bias Compensation Least Squares and the Alternate Algorithm. *Math. Probl. Eng.* **2020**, *2020*, 1DUMMMY. [CrossRef]
2. Schmalstieg, J.; Käbitz, S.; Ecker, M.; Sauer, D.U. A Holistic Aging Model for Li(NiMnCo)O_2 Based 18650 Lithium–Ion Batteries. *J. Power Sources* **2014**, *257*, 325–334. [CrossRef]
3. Broussely, M.; Herreyre, S.; Biensan, P.; Kasztejna, P.; Nechev, K.; Staniewicz, R.J. Aging Mechanism in Li Ion Cells and Calendar Life Predictions. *J. Power Sources* **2001**, *97–98*, 13–21. [CrossRef]
4. Sulzer, V.; Marquis, S.G.; Timms, R.; Robinson, M.; Chapman, S.J. Python Battery Mathematical Modelling (PyBaMM). *J. Open Res. Softw.* **2021**, *9*, 1–8. [CrossRef]
5. Yan, L.; Peng, J.; Gao, D.; Wu, Y.; Liu, Y.; Li, H.; Liu, W.; Huang, Z. A Hybrid Method with Cascaded Structure for Early–Stage Remaining Useful Life Prediction of Lithium–Ion Battery. *Energy* **2022**, *243*, 123038. [CrossRef]
6. Tang, X.; Liu, K.; Wang, X.; Liu, B.; Gao, F.; Widanage, W.D. Real–Time Aging Trajectory Prediction Using a Base Model–Oriented Gradient–Correction Particle Filter for Lithium–Ion Batteries. *J. Power Sources* **2019**, *440*, 227118. [CrossRef]
7. Xiong, W.; Xu, G.; Li, Y.; Zhang, F.; Ye, P.; Li, B. Early Prediction of Lithium–Ion Battery Cycle Life Based on Voltage–Capacity Discharge Curves. *J. Energy Storage* **2023**, *62*, 106790. [CrossRef]
8. Severson, K.A.; Attia, P.M.; Jin, N.; Perkins, N.; Jiang, B.; Yang, Z.; Chen, M.H.; Aykol, M.; Herring, P.K.; Fraggedakis, D.; et al. Data–Driven Prediction of Battery Cycle Life before Capacity Degradation. *Nat. Energy* **2019**, *4*, 383–391, ISBN 4156001903. [CrossRef]
9. Attia, P.M.; Severson, K.A.; Witmer, J.D. Statistical Learning for Accurate and Interpretable Battery Lifetime Prediction. *J. Electrochem. Soc.* **2021**, *168*, 090547. [CrossRef]
10. Li, X.; Wang, Z.; Yan, J. Prognostic Health Condition for Lithium Battery Using the Partial Incremental Capacity and Gaussian Process Regression. *J. Power Sources* **2019**, *421*, 56–67. [CrossRef]
11. Richardson, R.R.; Osborne, M.A.; Howey, D.A. Gaussian Process Regression for Forecasting Battery State of Health. *J. Power Sources* **2017**, *357*, 209–219. [CrossRef]
12. Gao, D.; Huang, M. Prediction of Remaining Useful Life of Lithium–Ion Battery Based on Multi–Kernel Support Vector Machine with Particle Swarm Optimization. *J. Power Electron.* **2017**, *17*, 1288–1297. [CrossRef]
13. Fei, Z.; Zhang, Z.; Yang, F.; Tsui, K.L.; Li, L. Early–Stage Lifetime Prediction for Lithium–Ion Batteries: A Deep Learning Framework Jointly Considering Machine–Learned and Handcrafted Data Features. *J. Energy Storage* **2022**, *52*, 104936. [CrossRef]
14. Zhang, Y.; Peng, Z.; Guan, Y.; Wu, L. Prognostics of Battery Cycle Life in the Early–Cycle Stage Based on Hybrid Model. *Energy* **2021**, *221*, 119901. [CrossRef]
15. Xu, F.; Yang, F.; Fei, Z.; Huang, Z.; Tsui, K.L. Life Prediction of Lithium–Ion Batteries Based on Stacked Denoising Autoencoders. *Reliab. Eng. Syst. Saf.* **2021**, *208*, 107396. [CrossRef]
16. Yao, J.; Powell, K.; Gao, T. A two-stage deep learning framework for early-stage lifetime prediction for lithium-ion batteries with consideration of features from multiple cycles. *Front. Energy Res.* **2020**, *10*, 1059126.
17. Hsu, C.W.; Xiong, R.; Chen, N.Y.; Li, J.; Tsou, N.T. Deep Neural Network Battery Life and Voltage Prediction by Using Data of One Cycle Only. *Appl. Energy* **2022**, *306*, 118134. [CrossRef]
18. Aykol, M.; Gopal, C.B.; Anapolsky, A.; Herring, P.K.; van Vlijmen, B.; Berliner, M.D.; Bazant, M.Z.; Braatz, R.D.; Chueh, W.C.; Storey, B.D. Perspective—Combining Physics and Machine Learning to Predict Battery Lifetime. *J. Electrochem. Soc.* **2021**, *168*, 030525. [CrossRef]
19. Thelen, A.; Lui, Y.H.; Shen, S.; Laflamme, S.; Hu, S.; Ye, H.; Hu, C. Integrating Physics–Based Modeling and Machine Learning for Degradation Diagnostics of Lithium–Ion Batteries. *Energy Storage Mater.* **2022**, *50*, 668–695. [CrossRef]

20. Najera–Flores, D.A.; Hu, Z.; Chadha, M.; Todd, M.D. A Physics–Constrained Bayesian Neural Network for Battery Remaining Useful Life Prediction. *Appl. Math. Model.* **2023**, *122*, 42–59. [CrossRef]
21. Shi, J.; Rivera, A.; Wu, D. Battery Health Management Using Physics–Informed Machine Learning: Online Degradation Modeling and Remaining Useful Life Prediction. *Mech. Syst. Signal Process.* **2022**, *179*, 109347. [CrossRef]
22. Deng, Z.; Lin, X.; Cai, J.; Hu, X. Battery Health Estimation with Degradation Pattern Recognition and Transfer Learning. *J. Power Sources* **2022**, *525*, 231027. [CrossRef]
23. Jiang, B.; Gent, W.E.; Mohr, F.; Ermon, S.; Chueh, W.C.; Richard, D.; Jiang, B.; Gent, W.E.; Mohr, F.; Das, S.; et al. Bayesian Learning for Rapid Prediction of Lithium–Ion Battery–Cycling Protocols of Lithium–Ion Battery–Cycling Protocols. *Joule* **2021**, *5*, 3187–3203. [CrossRef]
24. Chen, B.R.; Kunz, M.R.; Tanim, T.R.; Dufek, E.J. A Machine Learning Framework for Early Detection of Lithium Plating Combining Multiple Physics–Based Electrochemical Signatures. *Cell Reports Phys. Sci.* **2021**, *2*, 100352. [CrossRef]
25. Chen, B.R.; Walker, C.M.; Kim, S.; Kunz, M.R.; Tanim, T.R.; Dufek, E.J. Battery Aging Mode Identification across NMC Compositions and Designs Using Machine Learning. *Joule* **2022**, *6*, 2776–2793. [CrossRef]
26. Ansean, D.; Garcia, V.M.; Gonzalez, M.; Blanco–Viejo, C.; Viera, J.C.; Pulido, Y.F.; Sanchez, L. Lithium–Ion Battery Degradation Indicators Via Incremental Capacity Analysis. *IEEE Trans. Ind. Appl.* **2019**, *55*, 2992–3002. [CrossRef]
27. Ahmed, M.; Seraj, R.; Islam, S.M.S. The K–Means Algorithm: A Comprehensive Survey and Performance Evaluation. *Electron.* **2020**, *9*, 1295. [CrossRef]
28. Cortes, C.; Vapnik, V. Support–Vector Networks. *Mach. Learn.* **1995**, *20*, 273–297. [CrossRef]
29. Drucker, H.; Surges, C.J.C.; Kaufman, L.; Smola, A.; Vapnik, V. Support Vector Regression Machines. *Adv. Neural Inf. Process. Syst.* **1997**, *1*, 155–161.
30. Smola, A.J.; Schölkopf, B. A Tutorial on Support Vector Regression. *Stat. Comput.* **2004**, *14*, 199–222. [CrossRef]
31. Pinson, M.B.; Bazant, M.Z. Theory of SEI Formation in Rechargeable Batteries: Capacity Fade, Accelerated Aging and Lifetime Prediction. *J. Electrochem. Soc.* **2013**, *160*, A243–A250. [CrossRef]
32. Gao, T.; Han, Y.; Fraggedakis, D.; Das, S.; Zhou, T.; Yeh, C.N.; Xu, S.; Chueh, W.C.; Li, J.; Bazant, M.Z. Interplay of Lithium Intercalation and Plating on a Single Graphite Particle. *Joule* **2021**, *5*, 393–414. [CrossRef]
33. Chowdhury, S.; Lin, Y.; Liaw, B.; Kerby, L. Evaluation of Tree Based Regression over Multiple Linear Regression for Non–Normally Distributed Data in Battery Performance. In Proceedings of the 2022 International Conference on Intelligent Data Science Technologies and Applications (IDSTA), San Antonio, TX, USA, 5–7 September 2022; pp. 17–25. [CrossRef]
34. Tanim, T.R.; Paul, P.P.; Thampy, V.; Cao, C.; Steinrück, H.G.; Nelson Weker, J.; Toney, M.F.; Dufek, E.J.; Evans, M.C.; Jansen, A.N.; et al. Heterogeneous Behavior of Lithium Plating during Extreme Fast Charging. *Cell Rep. Phys. Sci.* **2020**, *1*, 100114. [CrossRef]
35. Tanim, T.R.; Dufek, E.J.; Evans, M.; Dickerson, C.; Jansen, A.N.; Polzin, B.J.; Dunlop, A.R.; Trask, S.E.; Jackman, R.; Bloom, I.; et al. Extreme Fast Charge Challenges for Lithium–Ion Battery: Variability and Positive Electrode Issues. *J. Electrochem. Soc.* **2019**, *166*, A1926–A1938. [CrossRef]
36. Gallagher, K.G.; Trask, S.E.; Bauer, C.; Woehrle, T.; Lux, S.F.; Tschech, M.; Lamp, P.; Polzin, B.J.; Ha, S.; Long, B.; et al. Optimizing Areal Capacities through Understanding the Limitations of Lithium–Ion Electrodes. *J. Electrochem. Soc.* **2016**, *163*, A138–A149. [CrossRef]
37. Tomaszewska, A.; Chu, Z.; Feng, X.; O'Kane, S.; Liu, X.; Chen, J.; Ji, C.; Endler, E.; Li, R.; Liu, L.; et al. Lithium–Ion Battery Fast Charging: A Review. *eTransportation* **2019**, *1*, 100011. [CrossRef]

batteries

MDPI

Article

An Evaluation Modeling Study of Thermal Runaway in Li-Ion Batteries Based on Operation Environments in an Energy Storage System

Min-Haeng Lee, Sung-Moon Choi, Kyung-Hwa Kim, Hyun-Sang You, Se-Jin Kim and Dae-Seok Rho *

Department of Electrical Engineering, Korea University of Technology & Education (KUT), Cheonan-si 31253, Republic of Korea; dlalsgod97@koreatech.ac.kr (M.-H.L.); moon9507@koreatech.ac.kr (S.-M.C.); kyunghwa316@koreatech.ac.kr (K.-H.K.); hsyu@koreatech.ac.kr (H.-S.Y.); tpwls1578@koreatech.ac.kr (S.-J.K.)
* Correspondence: dsrho@koreatech.ac.kr; Tel.: +82-41-560-1167

Abstract: According to the green growth and carbon-neutral policy in Korea, the installation of large-capacity ESSs is rapidly being increased, but a total number of 50 ESS fire cases have occurred since the end of 2023. ESSs are typically composed of series-parallel connections with numerous Li-ion batteries, and when the temperature of a deteriorated cell increases due to thermal, electrical, and mechanical stress, thermal runaway can occur due to additional heat generated by an internal chemical reaction. Here, an internal chemical reaction in a Li-ion battery results in the different characteristics on the decomposition reaction and heat release depending on the operation conditions in the ESS, such as the rising temperature rate, convective heat transfer coefficient, and C-rate of charging and discharging. Therefore, this paper presents mathematical equations and modeling of thermal runaway, composed of the heating device section, heat release section by chemical reaction, chemical reaction section at the SEI layer, chemical reaction section between the negative and positive electrodes and solvents, and chemical reaction section at the electrolyte by itself, based on MATLAB/SIMULINK (2022), which were validated by a thermal runaway test device. From the simulation and test results based on the proposed simulation modeling and test device according to the operation conditions in ESSs, it was found that the proposed modeling is an effective and reliable tool to evaluate the processing characteristics of thermal runaway because the occurrence time intervals and maximum temperatures had almost the same values in both the test device and simulation modeling. Accordingly, it was confirmed that the rising temperature rate and the convective heat transfer coefficient were more critical in the thermal runaway than the C-rate of charging and discharging.

Keywords: Li-ion battery; thermal runaway; operation environment; mathematical equation; evaluation modeling; MATLAB/SIMULINK

1. Introduction

Recently, the installation of large-capacity energy storage systems (ESSs) in South Korea have been rapidly increased to carry out various functions such as power stabilization of renewable energy sources, demand response, and frequency regulation, but the fire cases in ESSs have continuously occurred since August 2017 [1–3]. From the analysis results of 40 fire accident cases occurring from the end of 2022 based on the operation environments such as installation location, building type, and application, it is known that the fire accident rate in ESSs for renewable energy sources is much higher than other applications because they are installed and operate in relatively poor conditions [4,5]. On the other hand, ESSs are composed of series-parallel connections with a number of Li-ion batteries, and when the temperature of a deteriorated cell increases in a continuous way due to thermal, electrical, and mechanical stresses, a thermal runaway phenomenon can occur according to the additional heat generated by an internal chemical reaction [6–8]. Therefore, this paper formulated mathematical equations of thermal runaway in Li-ion batteries and then performed its evaluation modeling using MATLAB/SIMULINK (2022) S/W, which consists of a heating device section, a heat release section by chemical reaction, a chemical reaction section at the solid electrolyte interphase (SEI) layer, a chemical reaction section between negative and positive electrodes and solvents, and a chemical reaction section at the electrolyte by itself. Also, this paper implemented a test device for thermal runaway, which is composed of a heating chamber section, a heating device section, and a monitoring and control section in order to validate the effectiveness of the proposed modeling. From the simulation and test results based on the proposed simulation modeling and test device according to the operation conditions in ESSs, it is confirmed that the proposed modeling is an effective and reliable tool to evaluate the processing characteristics of thermal runaway because the occurrence time intervals and maximum temperatures in the test device are nearly identical to the results of simulation modeling. Also, it is found that as the convective heat transfer coefficient becomes higher due to the humidity and ventilation condition inside the battery module, the thermal runaway is occurs within a short time interval, and then the thermal runaway from a single cell can easily propagate to adjacent cells depending on the convective heat transfer coefficients. Based on the operation conditions in ESSs, it is clear that the rising temperature rate and the convective heat transfer coefficient are more critical in the thermal runaway than the C-rate of charging and discharging.

2. Mathematical Equations of Thermal Runaway in Li-Ion Batteries for ESSs

In order to evaluate the processing characteristics of thermal runaway in Li-ion batteries, this paper assumes that the temperature of a battery is raised by a heating chamber based on UL9450A, being one of the test methods shown in reference [9]. Here, the heat flux through the air in heating chamber is calculated by the convective heat transfer coefficient and the temperature difference between the heating device and battery, as shown in Equation (1) according to Newton's law of cooling. Also, the temperature variation in the battery due to the heating can be formulated as shown in Equation (2) by considering the heat flux and cross area, volume, density, and specific heat capacity of the current collectors in positive and negative electrodes [10,11].

$$q''(t) = k_{conv} \cdot (C_{heat} - C_{batt}(t)) \tag{1}$$

$$\Delta C_{batt,heat}(t) = \int_0^t \frac{2 \cdot D \cdot q''(t)}{v \cdot \rho_{batt} \cdot S_{heat}} dt \tag{2}$$

where q'' is the heat flux to the battery, k_{conv} is the convective heat transfer coefficient, C_{heat} is the temperature of the heating device, C_{batt} is the temperature of the battery, $\Delta C_{batt,heat}$ is the temperature variation in the battery due to heating, D is the cross area of the current collector in the battery, v is the volume of the current collector in the battery, ρ_{batt} is the density of the current collector in the battery, and S_{heat} is the specific heat capacity of the current collector in the battery.

Furthermore, the total heat release by internal chemical reactions occurring in a Li-ion battery can be expressed as shown in Equation (3), and the temperature variation in a battery due to chemical reactions is calculated as shown in Equation (4) [12–14].

$$Q_{total}(t) = Q_{sei} + Q_{nega} + Q_{posi} + Q_{elect} \tag{3}$$

$$\Delta C_{batt,chem}(t) = \int_0^t \frac{Q_{total}(t)}{\rho_{batt} \cdot S_{heat}} dt \tag{4}$$

where Q_{total} is the total heat release by internal chemical reactions, Q_{sei} is the heat release by the decomposition of the SEI layer, Q_{nega} is the heat release between the negative electrode and the solvent, Q_{posi} is the heat release between the positive electrode and the solvent, Q_{elect} is the heat release by the decomposition of the electrolyte, $\Delta C_{batt,chem}$ is the temperature variation in the battery due to the chemical reaction.

The heat release by the decomposition of the SEI layer can be formulated as shown in Equation (3), and the reaction velocity of the decomposition of the SEI layer is calculated as shown in Equations (4) and (5) according to the thermal decomposition reaction equation of a solid component based on the Arrhenius equation [15–17].

$$Q_{sei}(t) = H_{sei} \cdot \rho_c \cdot V_{sei}(t) \tag{5}$$

$$V_{sei}(t) = A_{sei} \cdot M_{sei}(t) \cdot exp^{\left(-\frac{e_{sei}}{R \cdot C_{batt}}\right)} \tag{6}$$

$$\frac{dM_{sei}(t)}{dt} = -V_{sei}(t) \tag{7}$$

where H_{sei} is the reaction heat by decomposition of the SEI layer, ρ_c is the volume-specific carbon content in the SEI layer, V_{sei} is the reaction velocity of the decomposition of the SEI layer, A_{sei} is the frequency factor of the decomposition of the SEI layer, e_{sei} is the activation energy of the decomposition of the SEI layer, R is the universal gas constant, and M_{sei} is the dimensionless amount of lithium-containing meta-stable species in the SEI layer.

In addition, the heat release between the negative electrode and solvent, which is a one-dimensional diffusion type, can be expressed as shown in Equation (8), and the reaction velocity between the negative electrode and solvent is calculated as shown in Equation (9). As the chemical reaction is continuously processed, the concentration of lithium in the negative electrode is decreased according to Equation (10), and the thickness of the SEI layer is increased based on Equation (11) due to the formation of the SEI layer.

$$Q_{nega}(t) = H_{nega} \cdot \rho_c \cdot V_{nega}(t) \tag{8}$$

$$V_{nega}(t) = -A_{nega} \cdot M_{nega}(t) \cdot exp^{\left(-\frac{W_{sei}(t)}{W_{sei}(0)}\right)} \cdot exp^{\left(-\frac{e_{nega}}{R \cdot C_{batt}}\right)} \tag{9}$$

$$\frac{dW_{sei}(t)}{dt} = V_{nega}(t) \tag{10}$$

$$\frac{dM_{nega}(t)}{dt} = -V_{nega}(t) \tag{11}$$

where H_{nega} is the reaction heat between the negative electrode and solvent, V_{nega} is the reaction velocity between the negative electrode and solvent, A_{nega} is the frequency factor between the negative electrode and solvent, W_{sei} is the thickness of the SEI layer, e_{nega} is the activation energy between the negative electrode and the solvent, and M_{nega} is the dimensionless amount of lithium within the carbon.

Also, the heat release between the positive electrode and solvent, which is an autocatalytic reaction, can be formulated as shown in Equation (12), and the reaction velocity between the positive electrode and solvent is calculated as shown in Equations (13) and (14).

$$Q_{posi}(t) = H_{posi} \cdot \rho_{posi} \cdot V_{posi}(t) \tag{12}$$

$$V_{posi}(t) = A_{posi} \cdot \alpha(t) \cdot (1 - \alpha(t)) \cdot exp^{\left(-\frac{e_{posi}}{R \cdot C_{batt}}\right)} \tag{13}$$

$$\frac{d\alpha(t)}{dt} = V_{posi}(t) \tag{14}$$

where H_{posi} is the reaction heat between the positive electrode and the solvent, ρ_{posi} is the volume-specific positive active content, V_{posi} is the reaction velocity between the positive electrode and the solvent, A_{posi} is the frequency factor between the positive electrode and the solvent, e_{posi} is the activation energy between the positive electrode and the solvent, and α is the conversion rate.

On the other hand, the heat release by decomposition of the electrolyte, which is a one-dimensional diffusion type, can be expressed as shown in Equation (15), and the reaction velocity of the decomposition of the electrolyte is calculated as shown in Equations (16) and (17).

$$Q_{elect}(t) = H_{elect} \cdot \rho_{elect} \cdot V_{elect}(t) \tag{15}$$

$$V_{elect}(t) = A_{elect} \cdot M_{elect}(t) \cdot exp^{\left(-\frac{e_{elect}}{R \cdot C_{batt}}\right)} \tag{16}$$

$$\frac{dM_{elect}(t)}{dt} = -V_{elect}(t) \tag{17}$$

where H_{elect} is the reaction heat by the decomposition of the electrolyte, ρ_{elect} is the volume-specific electrolyte active content, V_{elect} is the reaction velocity of the decomposition of the electrolyte, A_{elect} is the frequency factor of the decomposition of the electrolyte, e_{elect} is the activation energy of the decomposition of the electrolyte, and M_{elect} is the dimensionless concentration of the electrolyte.

Finally, this paper adapted the concept of a new single particle (SP) model found in reference [18] that includes electrolyte physics and a stress–diffusion coupling effect to describe the charging and discharging operation. Here, the analytical equations for electrolyte potential are derived based on the electrolyte charge conservation equation, and the stress-enhanced diffusion is coupled with electrochemical physics inside the entire battery [19,20]. Based on this concept, this paper presents the modeling of heat generation due to ohmic heat according to C-rate of charging and discharging by using the electrochemical–thermal coupled model as shown in reference [21].

3. Evaluation Modeling of Thermal Runaway Using MATLAB/SIMULINK S/W

Based on the proposed mathematical equations, the modeling of the heating device section can be expressed using MATLAB/SIMULINK, as shown in Figure 1. Here, Figure 1a shows the heat flux to the battery, which is created by Equation (1), and Figure 1b shows the temperature variation in the battery due to the heating, which is created by Equation (2).

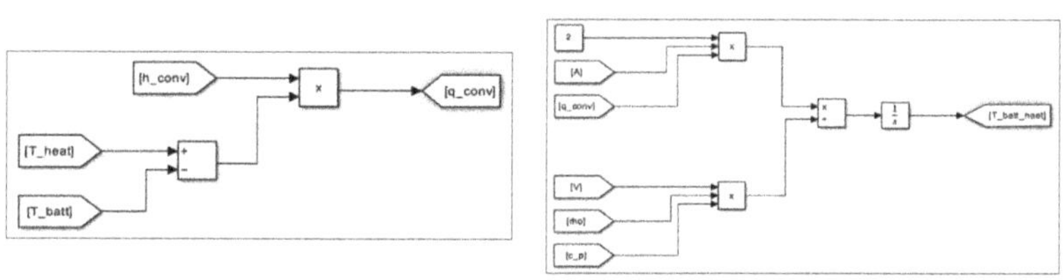

(a) (b)

Figure 1. Modeling of the heating device section for the battery: (**a**) heat flux to the battery; (**b**) temperature variation in the battery due to heating.

Also, the modeling of the heat release section by chemical reaction in a Li-ion battery can be illustrated as shown in Figure 2. Here, Figure 2a shows the total heat release by an internal chemical reaction, which is designed by Equation (3), and Figure 2b is the temperature variation of a battery due to internal chemical reactions, which is designed by Equation (4).

Furthermore, the modeling of the chemical reaction section at the SEI layer can be demonstrated as shown in Figure 3. Here, Figure 3a shows the modeling of heat release by the decomposition of the SEI layer, which is created by Equation (5), and Figure 3b is the reaction velocity of the decomposition of the SEI layer, which is created by Equations (6) and (7).

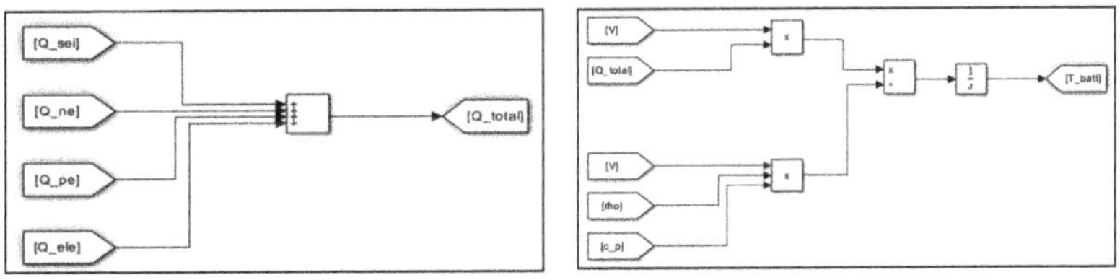

(a) (b)

Figure 2. Modeling of the heat release section by chemical reaction in a battery: (**a**) total heat release by internal chemical reactions; (**b**) temperature variation in the battery due to chemical reaction.

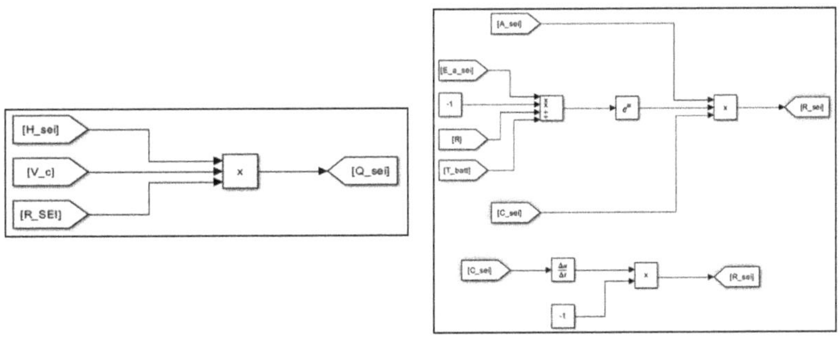

(a) (b)

Figure 3. Modeling of the chemical reaction section at the SEI layer: (**a**) heat release by the decomposition of the SEI layer; (**b**) reaction velocity of the decomposition of the SEI layer.

In addition, the chemical reaction section between the negative electrode and solvent can be expressed as shown in Figure 4. Here, Figure 4a shows the heat release between the negative electrode and the solvent, which is designed by Equation (8), and Figure 4b is the reaction velocity between the negative electrode and the solvent, which is designed by Equations (9)–(11). Also, the chemical reaction section between the positive electrode and solvent can be illustrated according to Equations (12)–(14) in the same way.

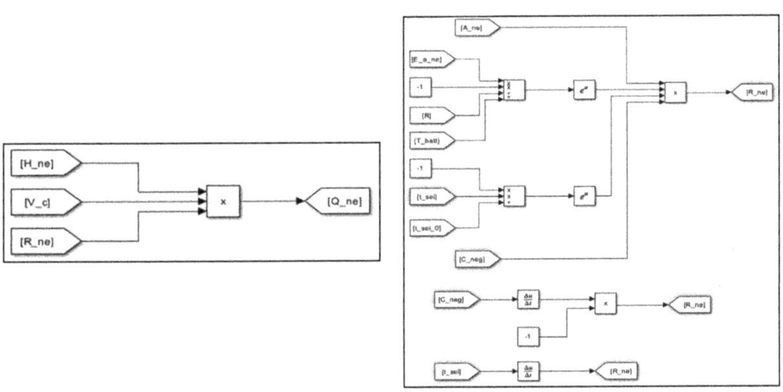

(a) (b)

Figure 4. Modeling of the chemical reaction section between the negative electrode and the solvent: (**a**) heat release between the negative electrode and the solvent; (**b**) reaction velocity between the negative electrode and the solvent.

Finally, the modeling of the chemical reaction section at the electrolyte by itself can be demonstrated as shown in Figure 5. Here, Figure 5a shows the heat release by the decomposition of the electrolyte, which is created by Equation (15), and Figure 5b is the reaction velocity of the decomposition of the electrolyte, which is created by Equations (16) and (17).

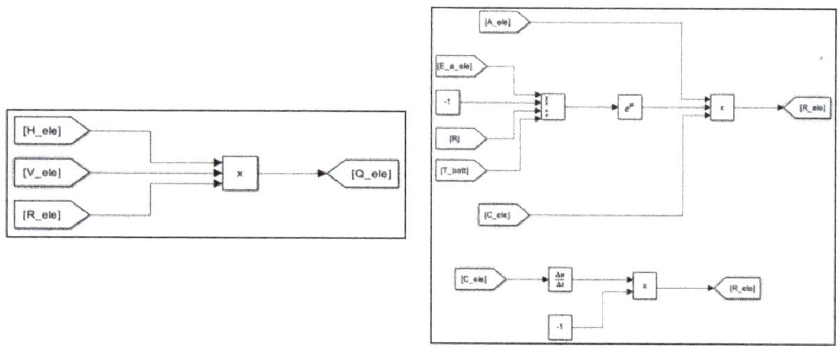

(a) (b)

Figure 5. Modeling of the chemical reaction section at the electrolyte: (**a**) heat release by the decomposition of the electrolyte; (**b**) reaction velocity of the decomposition of the electrolyte.

4. Implementation of a Test Device for Thermal Runaway in a Li-Ion Battery

In order to validate the proposed modeling, this paper designed and implemented the test device for thermal runaway in a Li-ion battery, which consists of a heating chamber section, a heating device section, and a monitoring and control section, and we evaluated the processing characteristics of thermal runaway due to external heating [22]. Here, the heating chamber section was assembled with reinforced steel frames to withstand thermal runaway, and a transparent polycarbonate window was installed on the front door to check the test processing. Also, the heating device and control sections were composed of a copper pipe, heat wire, and temperature controller, as shown in Figure 6. Here, a cylindrical Li-ion battery was inserted into the copper pipe, which was wrapped with heat wire, and then it was heated through the temperature controller to uniformly raise the temperature of the battery surface. In addition, the temperature of the battery was increased constantly by the RAMP function of the temperature controller, which was regulated accurately through temperature feedback from a thermocouple installed between the battery and copper pipe.

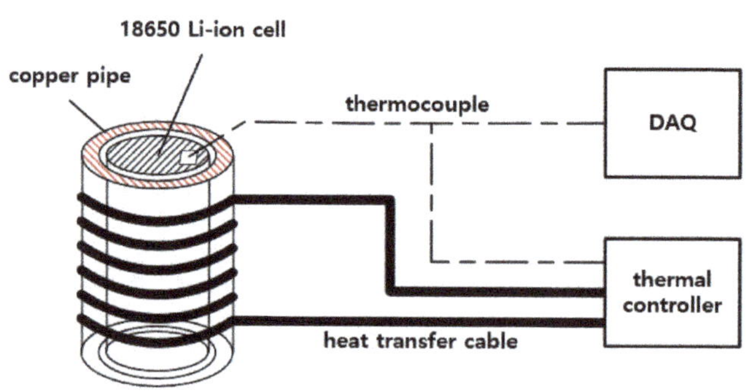

Figure 6. Configuration of the heating device.

Based on the components mentioned earlier, the outline of the implemented test device for thermal runaway in a Li-ion battery is shown in Figure 7. Here, Figure 7a shows the heating chamber section, and Figure 7b is the heating device and control section. Also, the thermal runaway test was performed in the laboratory with forced ventilation, explosion protection, and fire extinguishing systems to prevent an explosion caused by the flammable gases.

(**a**) (**b**)

Figure 7. Implementation of the test device for thermal runaway: (**a**) outline of the heating chamber section; (**b**) outline of the heating device and control section.

5. Case Studies

5.1. Simulation Conditions

In order to evaluate the processing characteristics of thermal runaway in a Li-ion battery according to operation environments, simulation conditions were assumed as shown in Table 1. Here, the rising temperature rates at the battery surface through the external heating were set as 3, 5, and 7 [°C/min], and the C-rates of charging and discharging were adopted as 0.5 and 5[C]. Also, the internal parameters in the 18650 cylindrical Li-ion battery were assumed as shown in Table 2 [23,24].

Table 1. Simulation conditions.

Items	Contents
Rising temperature rate [°C/min]	3, 5, 7
C-rate of charging and discharging [C]	0.5, 5
Convective heat transfer coefficient [W/(m^2·°C)]	6, 15

Table 2. Parameters of the internal Li-ion battery.

Symbols	Items	Values
R	universal gas constant	8.31
F	Faraday constant	96.485
H_{sei}	reaction heat by decomposition of the SEI layer	2.6×10^5
H_{nega}	reaction heat between the negative electrode and solvent	1.7×10^6
H_{posi}	reaction heat between the positive electrode and solvent	7.9×10^5
H_{elect}	reaction heat by the decomposition of the electrolyte	1.6×10^5
V_c	volume-specific carbon content in the SEI layer	1.7×10^3
V_{posi}	volume-specific positive active content	1.3×10^3
V_{elect}	volume-specific electrolyte active content	5×10^2

Table 2. *Cont.*

Symbols	Items	Values
A_{sei}	frequency factor of the decomposition of the SEI layer	1.7×10^{14}
A_{nega}	frequency factor between the negative electrode and solvent	5.0×10^{12}
A_{posi}	frequency factor between the positive electrode and solvent	2.3×10^{13}
A_{elect}	frequency factor of the decomposition of the electrolyte	5.1×10^{24}
e_{sei}	activation energy of the decomposition of the SEI layer	1.4×10^{5}
e_{nega}	activation energy between the negative electrode and solvent	1.4×10^{5}
e_{posi}	activation energy between the positive electrode and solvent	1.5×10^{5}
e_{elect}	activation energy of the decomposition of the electrolyte	2.7×10^{5}
M_{sei}	dimensionless amount of lithium-containing meta-stable species in the SEI layer	0.15
M_{nega}	dimensionless amount of lithium within the carbon in a negative electrode	0.75
M_{elect}	dimensionless concentration of the electrolyte	1
α	conversion rate	0.04
$W_{sei}(0)$	initial value of W_{sei}	0.033

5.2. Validation of the Proposed Modeling with Rising Temperature Rate

In order to validate the proposed modeling, the processing characteristics of thermal runaway in a Li-ion battery according to the test device and simulation modeling can be compared as shown in Figure 8. Here, the graphs ① and ② in Figure 8 show the thermal runaway results by using the test device and simulation modeling, in the case of the rising temperature rate of 5 [°C] per minute in the battery. Namely, the occurrence time intervals of thermal runaway by the test device and simulation modeling were obtained as 28 min 43 s and 29 min 19 s, respectively, and the maximum temperatures were 798 [°C] and 749.8 [°C]. Therefore, it is confirmed that the proposed modeling is an effective and reliable tool to evaluate the processing characteristics of thermal runaway because the occurrence time intervals and maximum temperatures had almost the same values in both the test device and the simulation modeling.

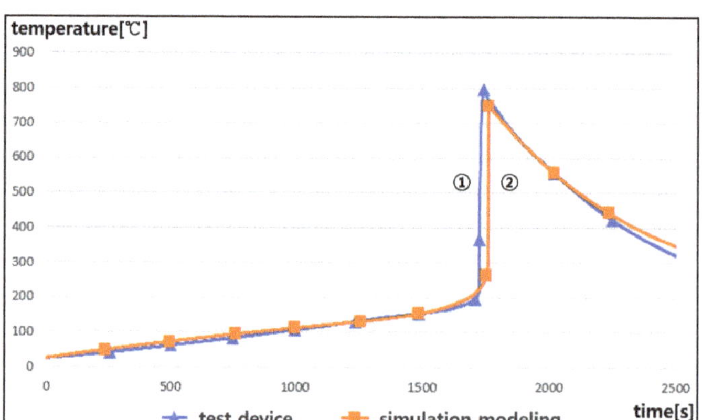

Figure 8. Profile of thermal runaway of the test device and simulation.

5.3. Thermal Runaway Characteristics with Rising Temperature Rate

Based on the simulation conditions mentioned above, the processing characteristics of thermal runaway according to the rising temperature rate were obtained as shown in Figure 9. As shown in the graph ① in Figure 9, when the rising temperature rate in the battery was by 3 [°C] per minute, thermal runaway occurred in about 46 min. Also, if the rising temperature rates were by 5 [°C] and 7 [°C] per minute, the occurrence time intervals of thermal runaway were calculated as 28 min and 20 min, respectively, as shown in the graphs ② and ③ in Figure 9. Comparing the graph ③ with graph ①, the time interval of thermal runaway was shortened by up to 26 min according to the rising temperature rate. Therefore, it was found that as the rising temperature rate became higher, the chemical reaction inside the battery was accelerated and then the occurrence time interval of thermal runaway was shortened.

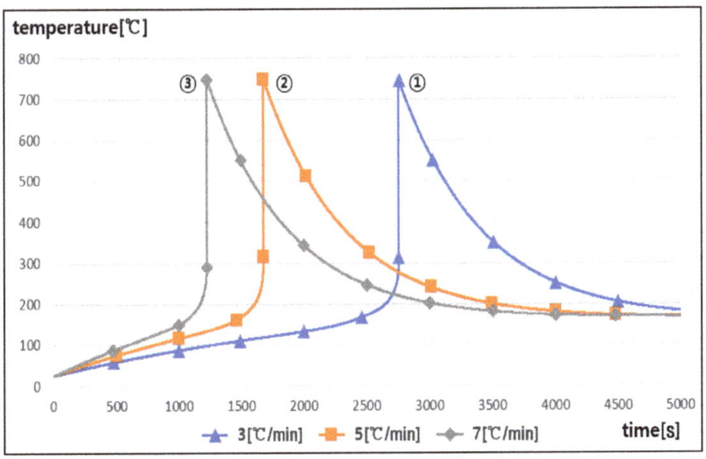

Figure 9. Profile of thermal runaway with the rising temperature rate.

5.4. Thermal Runaway Characteristics with the C-Rate of Charging and Discharging

The processing characteristics of thermal runaway in a Li-ion battery according to the C-rate of charging and discharging were obtained for the rising temperature rate of 5 [°C] per minute, as shown in Figure 10. The graph ① in Figure 10 shows the thermal runaway characteristics in the case of the 0.5 C-rate in charging and discharging, which applies for the test conditions in IEC 62619, and it was found that thermal runaway occurred in 29 min 20 s [25]. On the other hand, the graph ② in Figure 10 shows that the occurrence time interval of thermal runaway was calculated as about 28 min in the case of the 5 C-rate. Comparing the graph ② with graph ①, the time interval of thermal runaway was shortened by up to 80 s depending on the C-rate of charging and discharging. Consequently, it is clear that as the C-rate of charging and discharging became higher, the occurrence time interval of thermal runaway was slightly shortened due to the heat release from the internal resistance in the battery.

5.5. Thermal Runaway Characteristics with the Convective Heat Transfer Coefficient

The processing characteristics of thermal runaway in a Li-ion battery according to the convective heat transfer coefficient were obtained for the 270 [°C] of constant temperature in the heating chamber, as shown in Figure 11. The graph ① in Figure 11 shows the thermal runaway characteristics in the case of 6 [W/(m^2·°C)] of the convective heat transfer coefficient, which is the minimum value in air condition, and it is known that the temperature of the battery was increased by 5 [°C] per minute and thermal runaway occurred in about 29 min. On the other hand, the graph ② in Figure 11 shows the thermal runaway characteristics in the case of 15 [W/(m^2·°C)], which was made by forced ventilation of the battery module fan, and it was found that the temperature of the battery increased by 14.5 [°C] per minute and thermal runaway occurred in about 11 min. Namely, depending on the convective heat transfer coefficient, the temperature rise per minute was increased by about 9.5 [°C], and the time interval of thermal runaway was shortened by up to 18 min. Consequently, it is confirmed that as the convective heat transfer coefficient became higher due to the humidity and ventilation condition inside the battery module, the temperature of the battery rapidly rose and also thermal runaway occurred within a short time interval, and then thermal runaway from a single cell was easily able to propagate to adjacent cells depending on the convective heat transfer coefficients.

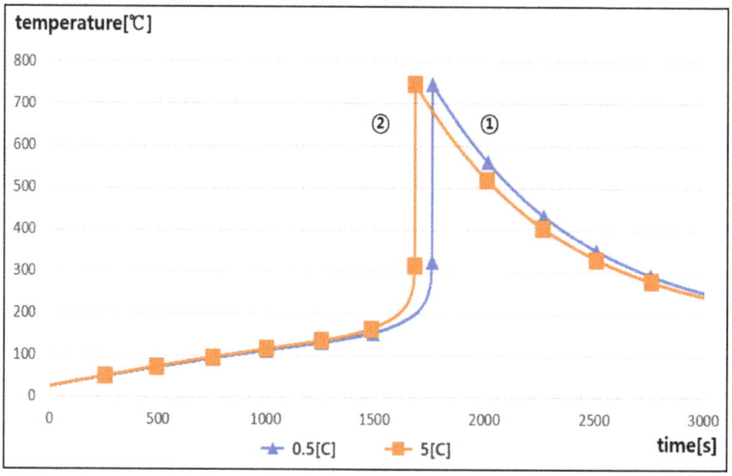

Figure 10. Profile of thermal runaway with the C-rate of charging and discharging.

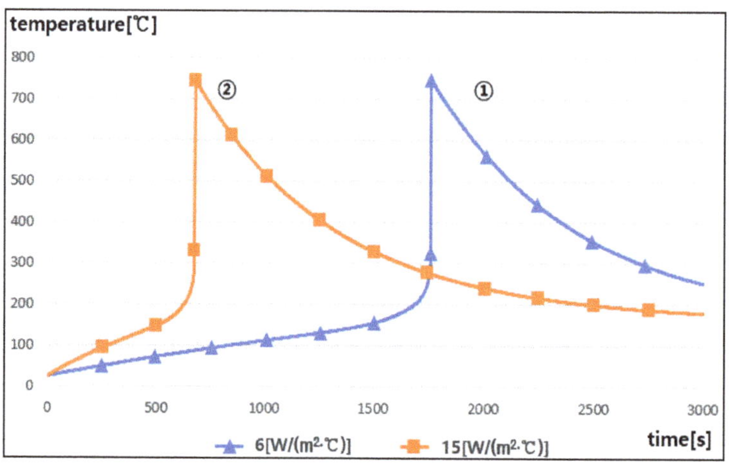

Figure 11. Profile of thermal runaway with the convective heat transfer coefficient.

5.6. Comprehensive Analysis

Based on the simulation results mentioned above, the processing characteristics of thermal runaway according to the operation conditions in ESSs such as the rising temperature rate, convective heat transfer coefficient, and C-rate of charging and discharging can be compared and summarized, as shown in Table 3. Based on the operation conditions, it is found that the rising temperature rate and the convective heat transfer coefficient were more critical in thermal runaway than the C-rate of charging and discharging. Therefore, in order to prevent thermal runaway in a Li-ion battery, it is necessary to install sensors monitoring the battery temperature and also cooling and ventilation systems that can cause suppression during the rapidly rising temperature. Also, an HVAC system should be applied to keep the optimal operation environment for ESSs.

Table 3. Occurrence time intervals of thermal runaway with the operation conditions in ESSs.

Operation Conditions		Occurrence Time Interval of Thermal Runaway [min]
Rising temperature rate [°C/min]	3	46
	5	28
	7	20
Convective heat transfer coefficient [W/(m^2·°C)]	6	29
	15	11
C-rate of charging and discharging [C]	0.5	29.33
	5	28

6. Conclusions

This paper has dealt with an evaluation modeling of thermal runaway using MATLAB/SIMULINK S/W based on the mathematical equations and it implemented the test device in order to evaluate the processing characteristics of thermal runaway according to operation conditions in ESSs. The main results of this paper are summarized as follows:

(1) It is confirmed that the proposed modeling is an effective and reliable tool to evaluate the processing characteristics of thermal runaway because the occurrence time intervals and maximum temperatures had almost the same values in both the test device and simulation modeling.

(2) It was found that as the rising temperature rate became higher, the chemical reaction inside the battery was accelerated and then the occurrence time interval of thermal runaway was shortened, and also, as the C-rate of charging and discharging became higher, the occurrence time interval of thermal runaway was slightly shortened due to the heat release from the internal resistance in the battery.

(3) It is confirmed that as the convective heat transfer coefficient became higher due to the humidity and ventilation condition inside the battery module, thermal runaway occurred within a short time interval, and then thermal runaway from a single cell can easily propagate to adjacent cells depending on the convective heat transfer coefficients.

(4) Based on the operation conditions in ESSs, it was found that the rising temperature rate and the convective heat transfer coefficient were more critical in thermal runaway than the C-rate of charging and discharging.

Author Contributions: All of the authors contributed to publishing this paper. M.-H.L., S.-M.C. and D.-S.R. carried out the test, modeling, and simulations, and they compiled the manuscript. The literature review and the simulation analysis were performed by K.-H.K., H.-S.Y. and S.-J.K., who also collected the data and investigated early works. All authors have read and agreed to the published version of the manuscript.

Funding: This research was supported by the Korea Institute of Energy Technology Evaluation and Planning (KETEP) grant funded by the Korean Government (MOTIE) (20224000000160, DC Grid Energy Innovation Research Center), and this research was supported the Korea Institute of Energy Technology Evaluation and Planning (KETEP) grant funded by the Korean Government (MOTIE) (RS-2024-00421994, Development of performance verification techniques and safety evaluation system for LiB-UPS System unit).

Data Availability Statement: The original contributions presented in the study are included in the article, further inquiries can be directed to the corresponding author.

Conflicts of Interest: The authors declare no conflicts of interest.

References

1. Nam, T.S. The here and the hereafter of Energy Storage System. *Trans. Korean Inst. Electr. Eng.* **2020**, *69*, 15–23.
2. Lim, S.Y.; Park, S.Y.; Yoo, S.H. The Economic Effects of the New and Renewable Energies Sector. *J. Energy Eng.* **2014**, *23*, 31–40. [CrossRef]
3. Park, K.M.; Kim, J.H.; Park, J.Y.; Bang, S.B. A Study on the Fire Risk of ESS through Fire Status and Field Investigation. *Fire Sci. Eng.* **2018**, *32*, 91–99. [CrossRef]
4. Lee, Y.B.; Kim, J.M.; Lee, M.H.; Rho, S.E.; Kim, S.J.; Rho, D.S. Safety Evaluation Method Considering Operation Environments and Applications in ESS. *Trans. Korean Inst. Electr. Eng.* **2024**, *73*, 773–783. [CrossRef]
5. Park, S.H.; Kang, J.G.; Kim, W.U.; Lim, H.W. A Study on Thermal Runaway Suppression Technology in Abnormal State for Energy Storage System(ESS) Using Lithium Secondary Battery. *J. Korean Inst. Illum. Electr. Install. Eng.* **2022**, *36*, 26–35.
6. Jang, H.J.; Song, T.S.; Kim, J.Y.; Kim, S.J.; Jang, T.H. Study on Analysis of Fire Factor and Development Direction of Standard/safety Requirement to Keep Safety for Energy Storage System (ESS). *Soc. Stand. Certif. Saf.* **2019**, *3*, 25–49. [CrossRef]
7. Jung, J.B.; Lim, M.G.; Kim, J.Y.; Rho, D.S. Characteristics of External Short-Circuit in Li-ion Battery Considering Operation and Environment Factors. *J. Korea Acad.-Ind. Coop. Soc.* **2021**, *22*, 663–672.
8. Lim, B.J.; Cho, S.H.; Lee, G.R.; Choi, S.M.; Park, C.D. Characteristics Analysis of Measurement Variables for Detecting Anomaly Signs of Thermal Runaway in Lithium-Ion Batteries. *Trans. Korean Hydrog. New Energy Soc.* **2022**, *33*, 85–94. [CrossRef]
9. *ANSI/CAN/UL 9540A(USA)*; Test Method for Evaluating Thermal Runaway Fire Propagation in Battery Energy Storage Systems. Sustainable Energy Action Committee: New York, NY, USA, 2019.
10. Peng, P.; Jiang, F. Thermal Safety of Lithium-ion Batteries with Various Cathode Materials: A Numerical Study. *Int. J. Heat Mass Transf.* **2016**, *103*, 1008–1016. [CrossRef]
11. Jiang, K.; Wang, T.; Li, X.; Duan, B.; Zhang, C. Simulation of Thermal Runaway Prediction Model for Nickel-rich Lithium ion Batteries. In Proceedings of the Chinese Automation Congress, Jinan, China, 20–22 October 2017; pp. 1293–1297.
12. An, Z.; Shah, K.; Jia, L.; Ma, Y. Modeling and Analysis of Thermal Runaway in Li-ion Cell. *Appl. Therm. Eng.* **2019**, *160*, 113960. [CrossRef]
13. Ren, D.; Feng, X.; Lu, L.; Ouyang, M.; Zheng, S.; Li, J.; He, X. An Electrochemical-thermal Coupled Overcharge-to-thermal-runaway Model for Lithium Ion Battery. *J. Power Sources* **2017**, *364*, 328–340. [CrossRef]
14. Liu, H.; Wei, Z.; He, W.; Zhao, J. Thermal issues about Li-ion batteries and recent progress in battery thermal management systems: A review. *Energy Convers. Manag.* **2017**, *150*, 304–330. [CrossRef]
15. Coman, P.T.; Darcy, E.C.; Veje, C.T.; White, R.E. Modelling Li-Ion Cell Thermal Runaway Triggered by an Internal Short Circuit Device Using an Efficiency Factor and Arrhenius Formulations. *J. Electrochem. Soc.* **2017**, *164*, 587–593. [CrossRef]
16. MacNeil, D.D.; Dahn, J.R. Test of Reaction Kinetics Using Both Differential Scanning and Accelerating Rate Calorimetries As Applied to the Reaction of Li_xCoO_2 in Non-aqueous Electrolyte. *Am. Chem. Soc.* **2001**, *105*, 4430–4439. [CrossRef]
17. Abada, S.; Marlair, G.; Lecocq, A.; Petit, M.; Sauvant-Moynot, V.; Huet, F. Safety focused modeling of lithium-ion batteries: A review. *J. Power Sources* **2016**, *306*, 178–192. [CrossRef]
18. Li, J.; Lotfi, N.; Landers, R.G.; Park, J. A Single Particle Model for Lithium-Ion Batteries with Electrolyte and Stress-Enhanced Diffusion Physics. *J. Electrochem. Soc.* **2017**, *164*, 874–883. [CrossRef]
19. Guduru, A.; Northrop, P.W.; Jain, S.; Crothers, A.C.; Marchant, T.R.; Subramanian, V.R. Analytical solution for electrolyte concentration distribution in lithium-ion batteries. *J. Appl. Electrochem.* **2012**, *42*, 189–199. [CrossRef]
20. Baba, N.; Yoshida, H.; Nagaoka, M.; Okuda, C.; Kawauchi, S. Numerical simulation of thermal behavior of lithium-ion secondary batteries using the enhanced single particle model. *J. Power Sources* **2014**, *252*, 214–228. [CrossRef]
21. An, Z.; Jia, L.; Wei, L.; Dang, C.; Peng, Q. Investigation on Lithium-ion Battery Electrochemical and Thermal Characteristic Based on Electrochemical-thermal Coupled Model. *Appl. Therm. Eng.* **2018**, *137*, 792–807. [CrossRef]
22. Hwang, S.Y.; Choi, S.M.; Jian, S.; Choi, H.S.; Rh, D.S. A Study on the Detection Algorithm of Off-gas to Prevent Thermal runaway of Li-ion Battery for ESS. *Trans. Korean Inst. Electr. Eng.* **2022**, *71*, 1787–1795. [CrossRef]
23. Kim, G.H.; Pesaran, A.; Spotnitz, R. A Three-dimensional Thermal Abuse Model for Lithium-ion Cells. *J. Power Sources* **2007**, *170*, 476–489. [CrossRef]

24. Wang, H.; Du, Z.; Rui, X.; Wang, S.; Jin, C.; He, L.; Zhang, F.; Wang, Q.; Feng, X. A Comparative Analysis on Thermal Runaway Behavior of Li ($Ni_xCo_yMn_z$) O_2 Battery with Different Nickel Contents at Cell and Module Level. *J. Hazard. Mater.* **2020**, *393*, 122361. [CrossRef] [PubMed]
25. *IEC 62619:2022*; Secondary Cells and Batteries Containing Alkaline or Other Non-Acid Electrolytes—Safety Requirements for Secondary Lithium Cells and Batteries, for Use in Industrial Applications. International Electrotechnical Commission: Geneva, Switzerland, 2022.

batteries

MDPI

Article

Linear Regression-Based Procedures for Extraction of Li-Ion Battery Equivalent Circuit Model Parameters

Vicentiu-Iulian Savu [1,*], Chris Brace [1], Georg Engel [2], Nico Didcock [2], Peter Wilson [1], Emre Kural [2] and Nic Zhang [1]

1 Faculty of Engineering and Design, University of Bath, Claverton Down, Bath BA2 7AY, UK; enscjb@bath.ac.uk (C.B.); prw30@bath.ac.uk (P.W.); qz254@bath.ac.uk (N.Z.)
2 AVL List GmbH, Hans-List-Platz 1, 8020 Graz, Austria; georg.engel@avl.com (G.E.); nico.didcock@avl.com (N.D.); emre.kural@avl.com (E.K.)
* Correspondence: vis21@bath.ac.uk

Abstract: Equivalent circuit models represent one of the most efficient virtual representations of battery systems, with numerous applications supporting the design of electric vehicles, such as powertrain evaluation, power electronics development, and model-based state estimation. Due to their popularity, their parameter extraction and model parametrization procedures present high interest within the research community, with novel approaches at an elementary level still being identified. This article introduces and compares in detail two novel parameter extraction methods based on the distinct application of least squares linear regression in relation to the autoregressive exogenous as well as the state-space equations of the double polarization equivalent circuit model in an iterative optimization-type manner. Following their application using experimental data obtained from an NCA Sony VTC6 cell, the results are benchmarked against a method employing differential evolution. The results indicate the least squares linear regression applied to the state-space format of the model as the best overall solution, providing excellent accuracy similar to the results of differential evolution, but averaging only 1.32% of the computational cost. In contrast, the same linear solver applied to the autoregressive exogenous format proves complementary characteristics by being the fastest process but presenting a penalty over the accuracy of the results.

Keywords: equivalent circuit model; lithium-ion battery; battery model parametrization; autoregressive exogenous model; least squares linear regression; optimization; electric vehicles

1. Introduction

1.1. Motivation and Challenges

Electric and hybrid electric vehicles, since they offer an affordable, more efficient, and potentially sustainable form of propulsion compared to conventional internal combustion vehicles, are receiving an increasing amount of attention from the research community. One of the defining elements of one vehicle is considered the battery system, as its development, manufacturing, and operation have significant implications over the lifespan of the vehicle, as well as its disposal and recycling. Consequently, research and development of elements contributing at each stage of the product cycle receive considerable attention as they have the potential to reduce cost while enhancing the performance and lifespan of the battery system.

One such element is represented by simulation models, which can range from simplistic static data-driven representations to complex multiscale and multidimensional internal representations of the electrochemical cell. Due to its attractive balance between the resources required to develop and the capability to replicate the battery system voltage response with sufficient accuracy, one of the most popular options is presented by equivalent circuit models (ECMs). These play a significant role during the development cycle, with applications such as simulating the battery using Hardware-in-the-Loop systems, powertrain evaluation, as well as the development of power electronics such as the battery management system (BMS). Contributions from this model are also present in the live management of the battery during vehicle operation, as model-based state estimation within the BMS is used to complete a series of functions, with the most popular being the estimation for the State of Charge (SOC).

1.2. Literature Review

Accurate parametrization represents one of the most important prerequisites for a model to be effective in any application, placing the extraction of parameter values from data as well as the structure of parameters under attention. In relation to ECMs, numerous methodologies are already available in specialized literature and cover a wide array of use cases. Summaries proposed by Kalogiannis et al. [1] and Li et al. [2] indicate the main techniques used in relation to time-domain data to be based on least squares linear regression (LS), Kalman filters (KF), generic optimizers, and analytical equations applied to step responses. If the model is parametrized in the frequency domain, electrochemical impedance spectroscopy is commonly employed to directly obtain the parameter values.

The essential parametrization procedures employing least squares linear regression are founded on the property of the ECM being a linear dynamic model. This implies the model can be formulated as a discrete transfer function and implicitly as an autoregressive exogenous model (ARX). This procedure is well covered by system identification theory, as highlighted by sources such as Ljung, L. [3], while in relation to battery ECMs, an exemplification of the simple procedure is presented by Zhang et al. [4]. More advanced studies, such as the one presented by Huang, C.-S. [5], target the improvement of the results obtained by this procedure from one data sample by considering characteristics such as the load profile, gradient of Open Circuit Voltage with respect to SOC, and condition number in the segmentation process of the data.

A more commonly employed extension of least squares applied to the ARX model procedure is presented by recursive least squares (RLS). The recursive part refers to a secondary component to the aforementioned procedure which recomputes the results and updates parameter values recursively. A plain exemplification of the process is presented by Raihan et al. [6], while most recent studies targeting this type of procedure attempt to increase the accuracy and robustness of the methodology through newly added elements. These include condition numbers as presented by Kim et al. [7], single or multiple forgetting factors, as suggested by Fan et al. [8] and Xia et al. [9], adapting forgetting factors, as exemplified by Sun, X. et al. [10], and computation over one specific window length (fixed memory), as demonstrated in the study by Sun, C. et al. [11]. In line with the fact that RLS-based studies generally discuss and target online applications, these benefit substantially live parameter estimation methodologies, as they increase the flexibility and ability of the procedure to track more accurately the change in parameters. However, the additions also present a series of downsides, such as the additional computational power as well as expertise and engineering effort to set up correctly. The latter is further highlighted by the proposed procedure by Xia et al. [9] containing Particle Swarm Optimization to determine the optimal forgetting factors added to an RLS process. Nevertheless, as most studies target to improve the iterative update procedure and not the essential least squares linear regression component, in the case of offline procedures, the existing research fails to bring significant advantages. This results from the complete dataset being readily available, allowing for the results to be considered in other manners such as collectively rather than iteratively, meaning the tracking ability may not be necessary.

The same issue is shared by the group of procedures based on Kalman filters or similar methodologies as they present a highly similar probability inference mechanism intended for supporting the tracking of the identified parameter values. The additional computational and engineering effort required to effectively set this part of the process will often render this type of option uncommon. This is supported by a close examination of the work presented by Li et al. [12] targeting to use this procedure to track only the slow dynamics of the system and the similar particle filter-based approach proposed by Xu et al. [13]. The Kalman filter-based procedures present as a common process used in relation to battery ECMs but represent an unpopular option for model parametrization. Instead, as exemplified by Fan et al. [8], Xia et al. [9], or Sun, C. et al. [11], they present a common option for SOC estimation tasks.

Outside the popular extension from linear regression to RLS, two other subgroups target the application of the least squares regression procedure directly to the state-space format of model equations and present interest for offline applications. Since the ECM presents as a linear dynamic model, the response to a step input can be summarized as a simple exponential function whose partial derivatives with respect to each model parameter can be easily computed. Consequently, a curve-fitting approach employing nonlinear least squares regression (NLS) becomes feasible, iteratively reducing error by using the local linearization of the objective function with respect to the targeted parameters. This type of approach is exemplified by Tian et al. [14], while an alternative implementation of NLS is also presented by Tran et al. [15] by directly employing the nlinfit function of MATLAB. The alternative implementation replaces the analytical computation of the derivatives with multiple evaluations of the objective function, becoming applicable outside step response data— at the cost, however, of computational effort. A completely different approach, limited also to step response, is represented by work presented by Chen, C. [16], which, through repeated integration and regrouping of parameters, obtains a new set of linear parameters to be directly identified by applying linear regression.

Another group of options is presented by the procedures based on predefined optimization packages. Following their flexibility, ease of implementation, and accuracy, these represent one of the most popular state-of-the-art options in recent published studies. Exemplifications are presented by Al Rafei et al. [17] proposing the ECM parametrization using the Genetic Algorithm complemented by the Taguchi Method, while Zhou et al. [18] introduce a hybrid coupling between particle swarm optimization and simulated annealing for the same task. Hybridization of the procedure towards the benefit of computational efficiency and accuracy can also be achieved in series, as presented by Ghoulam et al. [19], by employing, in sequence, the Fminsearch, Fmincon, and Genetic Algorithm found as readily available within the MatLab Optimization Toolbox. More comprehensive investigations, such as the ones proposed by Cheng, Y. S. [20] and Hou et al. [21], also present comparisons between multiple optimization algorithms in relation to the ECM parametrization problem.

Nevertheless, the benefits of the optimization-based procedures are also severely offset by their very high computational demand. This is directly associated with the number of degrees of freedom present in the optimization (number of parameters) as well as the absence of information over the optimization task other than the repeated evaluations of the objective function. While existent research may attempt to indicate one type of optimization to converge faster than another, the comparisons are highly dependent on multiple factors such as the data employed, model structure, and initialization, and cannot be used as a guarantee against a differently defined use case. At the same time, research attempting to add more information over the optimization task from a more effective source could present a more significant and permanent advantage in computational effort.

The last main group of procedures suitable for ECM parametrization support the analytical extraction from step response data, with examples such as the ones presented by Lee, S. et al. [22]. While highly effective, this type of procedure presents as the most rudimentary and is advised only for one-time procedures. The recommendation follows their easy implementation at the expense of human assistance and the use of a limited amount of information captured in a step response in the absence of multiple repetitions.

Machine learning techniques were also considered of interest for battery modeling. These have been used previously for a different category of models, capturing the complex relations between various factors and the state of the system empirically. Exemplification for this type of approach is presented by the study proposed by Lee, J. et al. [23], targeting the prediction of remaining useful life using artificial neural networks. By comparison, the effective extraction of ECM parameters from an individual segment of data represents a significantly different task. The ECM is a useful and simple algorithmic representation of the battery in its current state, which can be calibrated very effectively using the procedure detailed here. However, modeling the complex variation in ECM parameters as a function of the battery system state (e.g., SOC, SOH) may be a beneficial application of machine learning complementing the proposed novel procedures.

1.3. Main Contribution

This work sets to introduce, describe, and analyze in detail two novel methodologies, representing the main research contribution of the paper. These present the capability of effectively extracting ECM parameters from individual samples of battery voltage response data and are based on least squares linear regression (commonly referred to as ordinary least squares and abbreviated as LS throughout this work).

The first procedure is based on the well-known transformation to the ARX format, where the proposed improvement is the repetition of the process as part of a fully automated and consistent optimization of the data preprocessing parameters. This addition targets directly a reduction in MSE for the LS-based identification of parameter values. The second procedure applies the same linear solver to both linear and nonlinear parameters of the state-space format equations of the model. For this second option, the research contribution is presented by the new two-step procedure structure and the methodology for computing the local linearization for the nonlinear parameters. These enable the effective application of LS towards the extraction of ECM parameter values from any type of data. Finally, a procedure based on a global optimizer, namely differential evolution, is detailed, without any direct research contribution, but with the main goal to provide an accuracy benchmark for the results, following an endorsement of accuracy in previous research for optimization-based procedures. The additional comparison against a global optimizer is also set to highlight the methods strengths and weaknesses as well as provide a better understanding of the suitability of each method in relation to any application.

By contrast with the existing published research, the first procedure proposed attempts to improve the accuracy of the LS component applied to the ARX format found within system identification theory as well as part of the RLS procedures by integrating an additional optimization adjusting the preprocessing of the data. At the same time, the second procedure represents an updated version of multiple groups summarized in this section. By comparison with the LS-ARX procedures, however, it can apply the LS straight to the state-space format without being limited to step response data. This limitation is presented by some of the NLS procedures or procedures employing integration. The effective integration of gradient information and LS also leads to a large reduction in computational effort, with no penalty over accuracy, in comparison to optimization-based procedures. The procedure can also be considered a case of nested linear/nonlinear optimization. The example presented by Cleary et al. [24] represents a procedure with the most similar structure to date relative to the second procedure proposed in this study. Nevertheless, the novel proposed method in this study has the capability of using linear regression for both groups of parameters. The consequence is presented by a substantial computational advantage, while the prior published work uses particle swarm optimization for the nonlinear part, stating the use of linear regression for this task as an impossibility.

1.4. Article Organization

The remaining part of the paper is structured as follows: Section 2 introduces the model structure, mathematical foundations and methodologies of both parametrization procedures, as well as the methodology employed for the acquisition of experimental data. Section 3 contains an overview of the results, targeting an overview of the accuracy and efficiency of each procedure. Section 4 contains a detailed analysis of each new parametrization procedure proposed, while Section 5 follows with concluding remarks.

2. Materials and Methods

2.1. Model Structure

The equivalent circuit model structure, commonly referred to as the 2RC or 'double polarization' model, illustrated in Figure 1, has been selected to demonstrate all 3 parametrization procedures. The selection for the second order was based on popularity as well as a compromise between accuracy and complexity, as none of the 3 processes presents any limitation towards a different order of the RC model. The generic continuous state-space model has been summarized in two main equations rearranged according to the state-space format. Equation (1) resulted from directly applying Kirchhoff's second law to the main circuit loop. Equation (2) was obtained by employing both the first and second law of Kirchhoff for the RC element loops together with a definition of capacitive current; its derivation can be found in Appendix A.

$$V_{Out}(t) = V_{OC} - \sum_{i=1}^{n} R_i I_i(t) - R_0 I_0(t) \tag{1}$$

$$\frac{dI_i(t)}{dt} = \frac{-1}{\tau_i} I_i(t) + \frac{1}{\tau_i} I_0(t) \tag{2}$$

where $\tau_i = R_i C_i$ represents the parameters associated with RC element number i, n is the order of the RC model, I_i is the current associated with each RC element, R_0 is the series resistance, I_0 the input current, and V_{Out} is the voltage response of the model. To achieve discretization of the model, the forwards Euler method presented by Zeigler, B. [25] was employed. The process turned Equation (2) into Equation (3), following the derivation captured in Appendix B, while Equation (1) does not require any changes.

$$I_i(t_{k+1}) = \left(1 - \frac{\Delta t}{\tau_i}\right) I_i(t_k) + \frac{\Delta t}{\tau_i} I_0(t_k) \tag{3}$$

where t_k is the time in seconds at time step k and $\Delta t = t_{k+1} - t_k$.

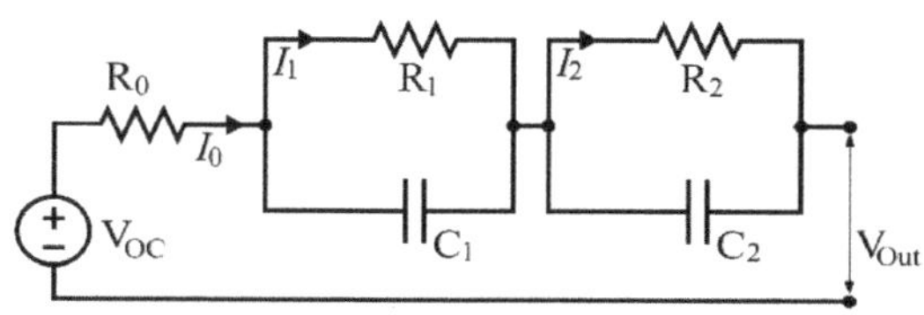

Figure 1. Double polarization ECM diagram.

To obtain the ARX format of the model captured by Equations (1) and (3), the z-transform process presented by Cheever, E. [26] is applied to both model equations. The result is then used to conduct the conversion from state-space form to a discrete transfer function, also documented by Cheever, E. [27]. The resulting model, having a constant term (c) to account for the Open Circuit Voltage parameter, is captured in Equation (4). The complete process described in this paragraph can be found in Appendix C. The equations defining the relations between ARX and state-space formats have also been summarized in Table 1 for the 2RC order model.

$$V_{Out}(z) = c - \sum_{i=1}^{n} a_i V_{Out} z^{-i} - \sum_{i=0}^{n} b_i I_0 z^{-i} \tag{4}$$

Table 1. ARX-state-space parameters mapping equations.

ARX Parameter	Equation
a_1	$\frac{\Delta t}{\tau_2} + \frac{\Delta t}{\tau_1} - 2$
a_2	$\frac{\Delta t^2}{\tau_1 \tau_2} - \frac{\Delta t}{\tau_1} - \frac{\Delta t}{\tau_2} + 1$
b_0	R_0
b_1	$\frac{R_0 \Delta t}{\tau_1} + \frac{R_0 \Delta t}{\tau_2} + \frac{R_1 \Delta t}{\tau_1} + \frac{R_2 \Delta t}{\tau_2} - 2R_0$
b_2	$\frac{R_0 \Delta t^2}{\tau_1 \tau_2} + \frac{R_1 \Delta t^2}{\tau_1 \tau_2} + \frac{R_2 \Delta t^2}{\tau_1 \tau_2} - \frac{R_0 \Delta t}{\tau_1} - \frac{R_0 \Delta t}{\tau_2} - \frac{R_1 \Delta t}{\tau_1} - \frac{R_2 \Delta t}{\tau_2} + R_0$
$c_{min/max}$	$(1 + a1 + a2) \times V_{OC,\,min/max}$

The study presented will only target individual segments of the experimental data. These have been assumed to present insignificant variation for parameter values caused by changes in operating conditions, namely SOC and temperature. Consequently, the parameters are assumed to maintain constant values throughout each segment of data, meaning only one locally correct value for each parameter to be identified. Modeling additional nonlinear dependence against temperature, SOC, or current amplitude for the parameter values is out of the scope of the work presented but represents the aim of future publications. Nevertheless, the assumption of one value being sufficient for each segment of data requires one exception for the Open Circuit Voltage (V_{OC}) parameter due to its significant variation relative to SOC. Consequently, two values are identified for each segment of data, associated with the minimum and maximum SOC values within each data sample. By assuming a local linear relation between the V_{OC} and SOC for the short SOC range covered by each data segment, the value at each time point is the result of linear interpolation between the two values. The change transforms Equations (1) and (4) into Equations (5) and (6), while the linear interpolation between the two distinct values is described by Equation (7). The exception is designed to enable the LS process to correctly attribute part of the dynamic behaviour to Open Circuit Voltage variation, which becomes significant, especially during long discharge pulses.

$$V_{Out}(t_{k+1}) = (V_{OC,max} \times p_{SOC}(t_k) + V_{OC,min} \times (1 - p_{SOC}(t_k))) - R_0 I_0(t_k) - \sum_{i=1}^{n} R_i I_i(t_k) \tag{5}$$

$$V_{Out}(z) = (c_{OCV,max} \times p_{SOC}(t_k) + c_{OCV,min} \times (1 - p_{SOC}(t_k))) - \sum_{i=1}^{n} a_i V_{Out} z^{-i} - \sum_{i=0}^{n} b_i I_0 z^{-i} \tag{6}$$

$$p_{SOC}(t_k) = \frac{SOC(t_k) - SOC_{min}}{SOC_{max} - SOC_{min}} \tag{7}$$

2.2. Parameter Extraction Procedures

The three proposed parameter extraction methods introduced are demonstrated using step response sections of data. Nevertheless, the mathematical underpinnings of all three allow extraction of parameters from any ad hoc measurements produced using an arbitrary current input to the battery system since both procedures do not include any assumption over the shape of the input current or voltage response.

The only assumption integrated targets the initial condition of the model, as the current through the resistance associated with each RC element is set to 0 at $k = 1$. This was considered reasonable as the cell was at the end of a long rest period before the current pulse was applied. The data employed were obtained from a Sony VTC6 and are part of a larger dataset acquired using the methodology described in Section 2.3. For direct comparison, the evaluation of results presented by the three procedures employed the same mean squared error (MSE) objective function described by Equation (8).

$$MSE = \frac{1}{m} \sum_{k=1}^{m} (V_{Exp}(t_k) - V_{Out}(t_k))^2 \tag{8}$$

where m is the number of time steps in the data sample, V_{Exp} is the experiment-measured voltage, and V_{Out} is the model response.

The objective for all three procedures is set to minimize the squared error between the model response and ad hoc time-domain experimental data. As a result, in the absence of further proof, the results should be considered an empirical fit over the data, and not one aligned with the physical attributes of the system capable of rendering a correct phenological/physical model. This observation carries importance, especially in the case of ad hoc data segments which may not capture sufficiently phenomena associated with the actual physical system. However, these can be used and replicated empirically by the ECM, with one of the most common examples being represented by drive cycle data.

2.2.1. Parameter Extraction Using the ARX Model Format and Least Squares Linear Regression (LS-ARX)

Following the description of linear dynamic systems using discrete transfer functions, rearranging the terms of the equation will result in what is commonly referred to as an autoregressive exogenous model (ARX). The identification of the model is well established in system identification theory and is achieved by employing LS, as demonstrated by Ljung, L. in his summary [3] and described in (9), which can also be solved by a linear solver. Once the ARX format parameters have been found, reversing the z-transform using the equations summarized in Table 1 will result in the state-space format parameter values.

$$\gamma = \left(A^T A\right)^{-1} A^T B \tag{9}$$

where $A,\ B$ are the design and response matrices for a data sample, with m time steps used to extract the parameters for the nth order of the model contained by matrix γ, as expanded below.

$$\gamma = \begin{bmatrix} c_{OCV,max} & c_{OCV,min} & a_1 & \cdots & a_n & b_0 & \cdots & b_n \end{bmatrix}^T$$

$$A = \begin{bmatrix} p_{soc}(t_{n+1}) & 1 - p_{soc}(t_{n+1}) & V_{Exp}(t_n) & \cdots & V_{Exp}(t_1) & I_0(t_{n+1}) & \cdots & I_0(t_1) \\ p_{soc}(t_{n+2}) & 1 - p_{soc}(t_{n+2}) & V_{Exp}(t_{n+1}) & \cdots & V_{Exp}(t_2) & I_0(t_{n+2}) & \cdots & I_0(t_2) \\ & & \ddots & & \ddots & \ddots & \ddots & \\ p_{soc}(t_m) & 1 - p_{soc}(t_m) & V_{Exp}(t_{m-1}) & \cdots & V_{Exp}(t_{m-n}) & I_0(t_m) & \cdots & I_0(t_{m-n}) \end{bmatrix}$$

$$B = \begin{bmatrix} V_{Exp}(t_{n+1}) \\ V_{Exp}(t_{n+2}) \\ \ddots \\ V_{Exp}(t_m) \end{bmatrix}$$

By considering the application of LS in the identification of ARX parameters from individual data segments, the significant effect of two factors has been described below, while also being supported by the results and discussion in Section 4.1.

Factor 1—conditioning of least squares linear regression. This factor reflects the variance of the LS result and is directly correlated with the observability of the parameters from the data provided as well as other mathematical concepts such as the covariance matrix and condition number. Examples of studies considering this factor are scarce and include studies such as the one presented by Kim et al. [7], which uses the condition number to set the forgetting factor in a recursive approach. A separate example is also provided by Huang, C.-S. [5], which uses multiple criteria to efficiently devise data samples. As a result, well-conditioning of the LS is rarely considered, despite having a significant impact on the results. The factors which can directly influence this are numerous, with the most common being the number of parameters to be identified (resulting from the order of the RC model selected and directly influencing the sensitivity of the model response relative to the parameters), the sampling of the data (high sample rates will reduce the fraction of the response captured by each time step), maneuvers captured by the data (lack of information), and behavior of the system (observability of individual phenomena in time-domain data). While improving the conditioning will likely result in more accurate results when applying LS, the true objective of the procedure to be presented is represented by the reduction in error between the experimental data and the simulated result. Consequently, if an additional process is set in support of the LS process, it should target directly the minimization of error rather than another indicator as the main objective. In the list of factors affecting the conditioning, the sampling rate and order of the model are the only controllable items when considering data which has already been acquired. Consequently, they are also the only ones that could be potentially optimized towards supporting a more accurate result.

Factor 2—error propagation. Noise is not considered a significant issue when considering the capability of LS to provide unbiased results for a problem. Nevertheless, when considering the application in relation to the ARX model, it can be observed that the voltage response values are captured not only by the response matrix but also by the design matrix of the regression. Consequently, the measurement noise associated with it will affect not only the dependent values of the regression but also the independent ones. This leads to a case of error-in-variables, which, in the case of simple LS, is set to cause a biased result, sometimes referred to as attenuation bias. This will cause an error margin correlated with the noise levels found within the experimental data.

The investigation presented in this paper will target to reduce the negative impact of the aforementioned issues associated with LS being applied to the ARX format model. By adding two additional degrees of freedom to the process, a direct reduction in error between the resulting model response and experimental data can be achieved. The degrees of freedom will control parameters used in the preprocessing of each data segment, namely the size of a moving average filter window as well as the sampling period, by introducing downsampling. Manually adjusting the same parameters may also represent an effective process in the case of a small number of data segments being processed by an experienced engineer. However, this option was not deemed a feasible approach in the case of large datasets or when automation of the process over multiple arbitrary datasets (e.g., experiment, real driving cycles) is desirable. Consequently, an automated search procedure is proposed to complement the extraction of the ARX parameters using LS to the benefit of both online and offline procedures.

The complete process is illustrated in Figure 2. The starting point is presented by the MSE value obtained for one data sample without conducting any preprocessing of the experimental data. This sets a benchmark result, which is then compared and potentially replaced with other results after the same data has been filtered using a moving average (MA) filter and sampled down to a lower data point frequency.

The control is achieved by 4 distinct values: 2 memorizing the current values and 2 memorizing the next values to be assessed for the filter window size (FWS) and the sampling period (SP) of the data. At the beginning of the process, the current FWS and sampling period are set to 1 and the default equidistant SP of the data, while the next values are set to 4 and double the value of the default SP. At each iteration, the algorithm will try to obtain a better result by using the next FWS value and the current value for the SP. If this results in a lower mean squared error (MSE), the next value becomes the current one, and the next FWS is multiplied by 2. However, if unsuccessful, downsampling the data by a higher factor is tried, hence using the current value for the FWS and the next value for the SP, followed by the same MSE test. This decides whether to set the current SP to the value of the next SP and multiply the latter by 2. If both attempts fail to bring an improvement, a failed iteration is counted and both next values are multiplied by 4, targeting the acceleration of the algorithm and potential avoidance of a local minimum. The algorithm allows for a maximum of 3 failed iterations, then stops and outputs the parameter values associated with the lowest mean squared error. The final result is represented by an optimization-type approach complementing the identification of ARX parameters using linear regression, a process which has not been previously presented in published research.

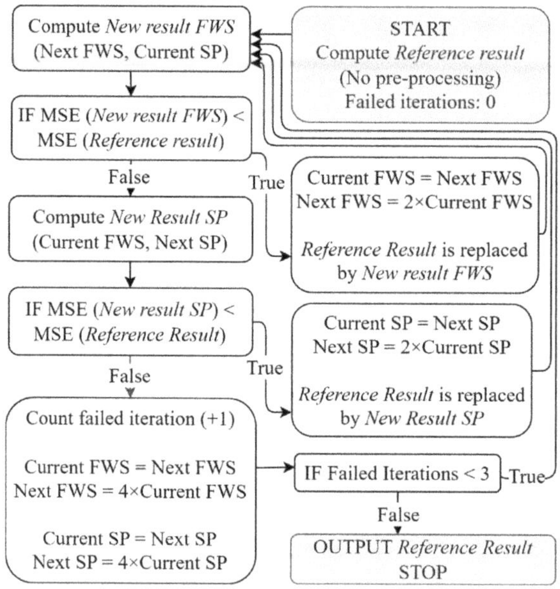

Figure 2. LS-ARX process flow diagram (FWS—filter window size; SP—sampling period).

The additional process described in support of least squares linear regression can also be interpreted as a search algorithm for the two-parameter values associated with the local minimum of the MSE via an iterative approach. This observation carries importance, as it also highlights that the procedure is not intended to apply filtering and downsampling procedures analytically following analysis of the frequency response of the system and noise characteristics. Instead, it will only seek alteration of the data segment and, implicitly, the results of the regression towards the benefit of the empirical reproduction of the system response.

The logic of the process attempting to alter the filtering process first during each iteration followed by the sampling period represents only a suggestion set on the assumption that noise within the data occurs more commonly than oversampling. The order is also supported by the fact that the moving average filter has the benefit of not completely dismissing information by removing data points by comparison with the downsampling process. Nevertheless, the search procedure can be configured to run in a different order, control a different preprocessing procedure, or be completely replaced with a readily available optimization or brute-force approach.

For completeness, the methodology does not require any initial values for the model parameters, while the objective function used to evaluate each result is the MSE defined by Equation (8). Due to the structure of the ARX model, the data used for parameter extraction need to be perfectly equidistant along its time axis, hence an additional function to fulfill this requirement must be implemented. An additional function is also required to convert the ARX to state-space parameter values by inverting the equations in Table 1.

2.2.2. Parameter Extraction Based on Linearization and Least Squares Linear Regression (LS-ECM)

Following the ECM in discrete state-space format being summarized in Equations (1) and (3), the parameters proposed for extraction are grouped into linear (open circuit voltage, resistances) and nonlinear (time constants) in relation to the voltage response of the model. Following this observation, a simple, more elegant and efficient solution previously not proposed in the literature in relation to battery system ECMs can be also constructed. This presents two stages, repeated iteratively, each targeting one group of parameters. The first step determines the linear values through the direct application of the linear solver. The design matrix of the step is obtained by using the derivative of the voltage response with respect to the linear parameters obtained from Equation (5) and expressed in Equations (10)–(12).

$$\frac{dV_{out}(t_k)}{dV_{OC,max}} = p_{SOC}(t_k) \quad (10)$$

$$\frac{dV_{out}(t_k)}{dV_{OC,min}} = 1 - p_{SOC}(t_k) \quad (11)$$

$$\frac{dV_{out}(t_k)}{dR_i} = I_i(t_k) \quad (12)$$

The second, novel stage follows a similar logic as the same linear solver will be used to optimize the values of the time constants. Since these are nonlinear parameters, local linearizations and multiple iterations are required to obtain a result, as the linear approximation is only locally viable. Nevertheless, the linearization of the nonlinear time constants will not be achieved through repetitive evaluation of the objective function. Instead, a more effective process is set by computing the total derivative of the voltage response with respect to the time constants by using the simulated data from the computation of the ECM after step 1.

The total derivative is computed by considering Equations (1) and (3) describing the discrete state-space second-order ECM. By applying the chain rule, the total derivative of voltage with respect to a time constant is expressed in Equation (13) as the sum of the partial derivative of voltage with respect to time constants (equal to 0) and the partial derivative of dependent terms (the current associated with each RC element) multiplied by their total derivative with respect to the time constant.

$$\frac{dV_{out}(t_k)}{d\tau_i} = \frac{\partial V_{out}(t_k)}{\partial \tau_i} + \frac{\partial V_{out}(t_k)}{\partial I_i(t_k)} \times \frac{dI_i(t_k)}{d\tau_i} \quad (13)$$

where $\frac{\partial V_{out}(t_k)}{\partial \tau_i} = 0$ and $\frac{\partial V_{out}(t_k)}{\partial I_i(t_k)} = R_i$, resulting in (14).

$$\frac{dV_{out}(t_k)}{d\tau_i} = R_i \times \frac{dI_i(t_k)}{d\tau_i} \quad (14)$$

The total derivative of the current associated with each RC element with respect to the time constant of the RC element is computed at each time step recursively using Equation (15), obtained by applying the chain rule again. Consequently, the total derivative of a current associated with an RC element with respect to the time constant at each time step is dependent on the same derivative of the previous time step. By following the assumption of no current flow through any resistance at $k = 1$, the first value in the chain of the total derivative can be set to 0.

$$\frac{dI_i(t_k)}{d\tau_i} = \frac{\partial I_i(t_k)}{\partial \tau_i} + \frac{\partial I_i(t_k)}{\partial I_i(t_{k-1})} \times \frac{dI_i(t_{k-1})}{d\tau_i} \quad (15)$$

The use of the total derivatives for identifying the nonlinear time constants has been proposed in the literature before for ECMs using linearization within NLS procedures, but always either limited to step response or computing the derivatives by evaluations of the objective function. In the absence of assumptions regarding the data, the extended simple or dual Kalman filter-based identification methods also use the same total derivative to adjust the values of the time constants. The novel contribution of the study is found in the second step of the process presented and is represented by the incorporation of the total derivatives determined analytically from data provided by the last model computation to implement an effective iterative optimization procedure for the time constants using a linear solver.

To apply the working principle described above, the iterative structure in Figure 3 using two groups of model parameters (linear—γ_1 and nonlinear—γ_2) has been constructed. During each iteration, in step 1, the second group of parameters is kept fixed, while the first group is adjusted using the linear solver, followed by an evaluation of the objective function, which is also used to compute the new values for the derivatives. This is repeated in step 2 for the adjustment of the second group, during which the first group is fixed, followed again by evaluation. Each iteration will use the linear solver to compute new values for the parameters considered in the first group and a correction for the parameters in the second group based on the error margin. For the parameters associated with the RC elements, additional boundaries are added in each iteration to improve the stability of the process, limiting the maximum correction to half of the current value. This prevents overcorrections caused by the difference between the linear approximation and the actual nonlinear objective function relative to the second group of parameters.

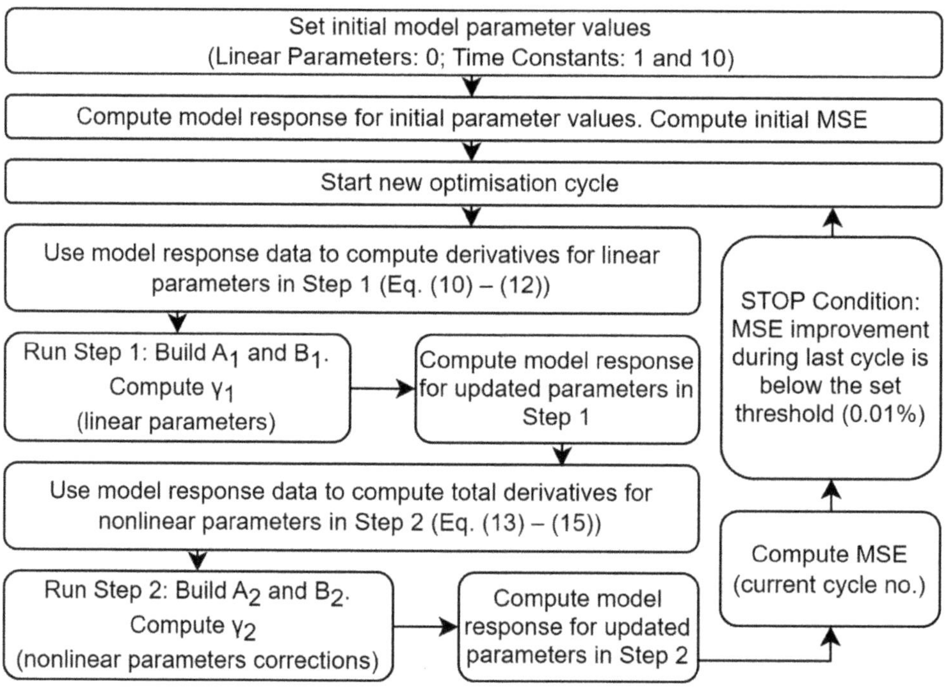

Figure 3. LS-ECM process flow diagram.

For completeness, the design and response matrices used in the application of the linear solver for both steps of the procedure have been expanded below. The design matrix is composed of the linear coefficients of each parameter at each time step in the data, namely the total derivative of voltage with respect to each parameter at each time step. The result is computed using the SciPy 1.5.4 linear solver named (lsq_linear) [28] with the default settings, while the same result could also be obtained by applying Equation (9).

- Step 1

$$\gamma_1 = \begin{bmatrix} V_{OC,min} & V_{OC,max} & R_0 & R_1 & \ldots & R_n \end{bmatrix}$$

$$A_1 = \begin{bmatrix} \frac{dV_{out}(t_1)}{dV_{OC,min}} & \frac{dV_{out}(t_1)}{dV_{OC,max}} & \frac{dV_{out}(t_1)}{dR_0} & \cdots & \frac{dV_{out}(t_1)}{dR_n} \\ \frac{dV_{out}(t_2)}{dV_{OC,min}} & \frac{dV_{out}(t_2)}{dV_{OC,max}} & \frac{dV_{out}(t_2)}{dR_0} & \cdots & \frac{dV_{out}(t_2)}{dR_n} \\ & \ddots & & \ddots & \\ \frac{dV_{out}(t_m)}{dV_{OC,min}} & \frac{dV_{out}(t_m)}{dV_{OC,max}} & \frac{dV_{out}(t_m)}{dR_0} & \cdots & \frac{dV_{out}(t_m)}{dR_n} \end{bmatrix} \quad B_1 = \begin{bmatrix} V_{Exp}(t_1) \\ V_{Exp}(t_2) \\ \ddots \\ V_{Exp}(t_m) \end{bmatrix}$$

- Step 2

$$\gamma_2 = \begin{bmatrix} \Delta\tau_1 & \cdots & \Delta\tau_n \end{bmatrix}$$

$$A_2 = \begin{bmatrix} \frac{dV_{out}(t_1)}{d\tau_1} & \cdots & \frac{dV_{out}(t_1)}{d\tau_n} \\ \frac{dV_{out}(t_2)}{d\tau_1} & \cdots & \frac{dV_{out}(t_2)}{d\tau_n} \\ \ddots & & \ddots \\ \frac{dV_{out}(t_m)}{d\tau_1} & \cdots & \frac{dV_{out}(t_m)}{d\tau_n} \end{bmatrix} \quad B_2 = \begin{bmatrix} (V_{Exp}(t_1) - V_{Out}(t_1)) \\ (V_{Exp}(t_2) - V_{Out}(t_2)) \\ \ddots \\ (V_{Exp}(t_m) - V_{Out}(t_m)) \end{bmatrix}$$

For completeness, the LS-ECM procedure requires a set of initial values for the parameters. These can be set to 0 for linear parameters and distinct, non-zero values for the time constants (in the examples presented, these were set to 1 and 10, following further discussion presented in relation to the results). The objective function is represented by the MSE defined by Equation (8), while the condition for the process to stop is an improvement in the MSE of less than 0.01% for one iteration, set subjectively to balance accuracy and efficiency. The requirement for equidistant data in the time axis is present also for this procedure; however, the accuracy provided by the raw data was considered satisfactory, as confirmed by the study results.

2.2.3. Extraction Based on Differential Evolution (DE-ECM)

For comparison purposes, a third process was set based on global optimization, without any alterations. The selection followed the popularity of this type of process in recent research and its potential to avoid a local minimum result in the absence of additional knowledge regarding the objective function shape. Additional motivation for the selection follows the global optimization-based procedures being considered a good standard for reliability and accuracy. All the parameters proposed for identification are set as variables for optimization and boundary values were required instead of initial values. These were set between 0 and a positive value, 10 for the Open Circuit Voltage, 1 for the resistance values, and 100 and 1000 for the time constants. By contrast with the proposed two methods, which present as deterministic, this process presents stochastic behavior. Consequently, for each sample, 10 separate runs were computed, with their average presented in the results section. The objective function used to drive the process is the same used for the LS-ARX and LS-ECM methods, while the readily available global optimizer was selected from the SciPy 1.5.4 library, namely the differential evolution (DE) algorithm [29]. The main role of this third method is to provide a representative benchmark for global optimization and a reliability standard, while also highlighting the efficiency of the method for applications where the model equations are ignored and not used to inform the parametrization method.

2.3. Experimental

The methods proposed for comparison were subject to assessment using experimental data. While the dataset obtained captures step discharges at SOC from 100% to 5% for 9 different temperatures, the dataset was not designed specifically for this activity, but also for more comprehensive modeling tasks. Also, despite the step response not being representative of any in-vehicle operation conditions, this format represents a very common type of data used for parameter extraction and implicit model parametrization following well-established system identification procedures. For the results presented in this work, a list of 3 short step responses (10 s) and 3 long constant current discharge steps (approximately 334 s) were selected. The resulting set was considered sufficient to demonstrate the working principle as well as compare the accuracy and efficiency of the processes.

The data were obtained from a pristine 3000 mAh NCA Sony VTC6 cell placed in an 18650 cell holder, as illustrated in Figure 4. The complete experimental setup displaying all the connections to the cell is also captured in Figure 5. To avoid imperfect contact between the plates of the holder and the cell, silver conductive was coated on contact surfaces before assembly. The cell was connected to equipment by having one set of wires (green and yellow) soldered to the plates of the holder on the opposite face to the contact surface with the cell to draw current, while wires intended to measure voltage (red/black) were soldered to the plates next to the point of contact with the cell, in an attempt to minimize overlap of the two circuits.

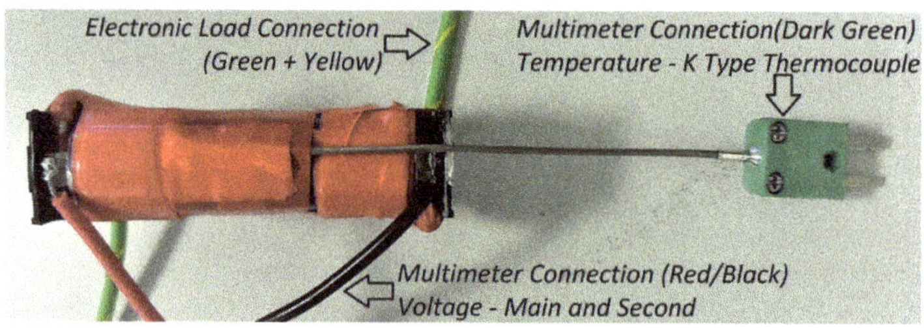

Figure 4. Sony VTC6 cell setup.

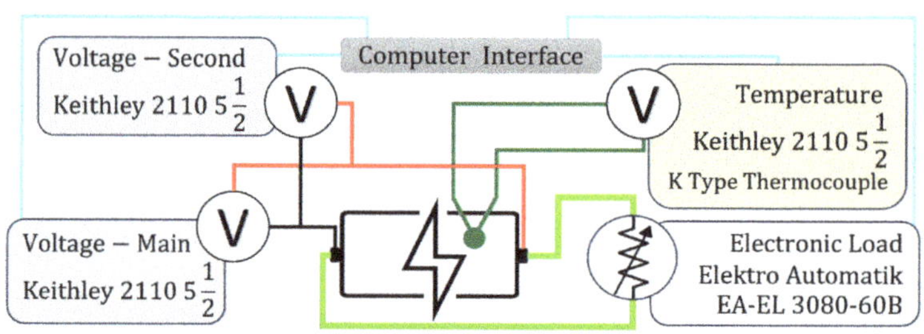

Figure 5. Experimental setup diagram.

The cell was connected to a programmable electronic load manufactured by EA Elektro-Automatik GmbH & Co. KG, Viersen, Germany, model EA-EL 3080-60 B, to accurately control the discharge current. The second set of wires was connected to two multimeters set in parallel, manufactured by Tektronix, Inc., Beaveron, OR, USA, model Keithley 2110 5 ½; this double voltage measurement setup was selected to avoid data loss in the case of a malfunction. The electronic load was controlled using the manual built-in interface, while the multimeters could feed the measured data directly to a computer. The computer interface for the multimeters was built as a Python 3.9.13 script based on Py-VISA 1.13.0 [30] and developed specifically for this activity. Lastly, a third multimeter was connected to a K-type thermocouple attached to the surface of the cell using thermally conductive adhesive and electrical tape, with the purpose of monitoring and controlling the temperature of the cell as well as recording it. Highly precise control of the sampling rate was not possible with the equipment available, either with the Python script developed or by using the proprietary software available for the multimeters. By considering the penalty in accuracy, the sample rate of the data was at approximately 20 Hz for the voltage measurements, while it was set to approximately 1.6 Hz for temperature. Further details about thermal control and monitoring are not considered relevant in this study as the data used were obtained in ambient conditions considered to be 20 °C.

The structure of one test at one specific temperature point is summarized in Figure 6 and follows charging the cell to 100%, defined as a 20 mA cut-off current at a constant voltage of 4.2 V, defined by the manufacturer in ambient conditions (20 °C). The charging is followed by a relaxation period of 30 min and another constant voltage charging to the same cut-off current value, which is added to avoid false SOC points due to hysteresis. This is followed by another relaxation period of a minimum of 30 min while the cell was conditioned to the correct temperature.

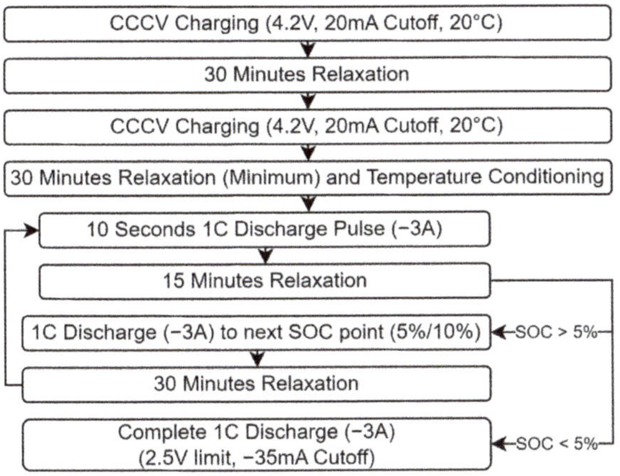

Figure 6. Battery cell data acquisition procedure flow diagram.

The test is commenced by a 10 s pulse discharge at 1C (−3A) associated with 100% SOC, which is then followed by 15 min of relaxation. The SOC points are then changed by 5% and 10% increments to the following values: 95, 85, 75, 65, 55, 45, 35, 25, 15, 10, and 5, once again using constant current at the same value of 1C. After discharging to the SOC point, a minimum period of 30 min of relaxation is allowed, after which the 10 s pulse is applied again, followed by 15 min of relaxation. After the last SOC point is observed under the pulse, the cell is completely discharged to 0%, defined as a −35 mA cut-off current at a constant voltage of 2.5 V. Through coulomb counting, this complete discharge is used to confirm the SOC breakpoints in the time series. Exemplification of the data employed is presented by the time seriesdisplayed in Figure 7. Segments numbered 1 to 3 (S1 to S3) capture individual segments associated with 10 s pulses, while segments 4 to 6 (S4 to S6) capture 10% SOC discharge steps.

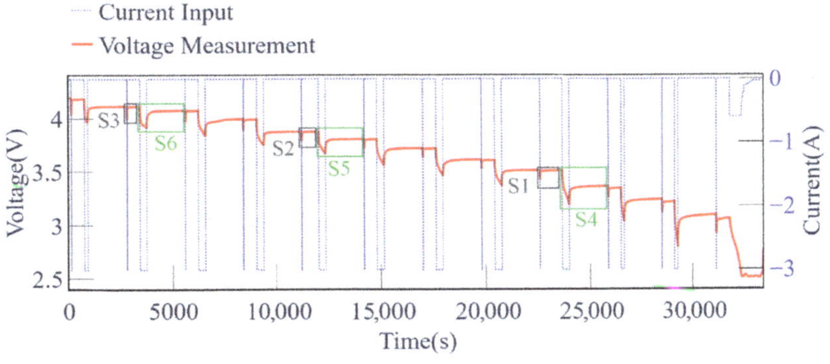

Figure 7. Sony VTC6 voltage response data (20 °C).

3. Results

The accuracy and efficiency of each method represent the first main observations. Table 2 summarizes the mean squared error (MSE) achieved by each of the methods proposed. A comparison of the measured data against the simulated response as well as the profile of the mean absolute error (MAE) have been presented for samples 1 and 4 in Figures 8 and 9 to further illustrate the results of the process, with highlights over regions of interest. The inclusion of only one result for each type of sample followed the similarity of the results across the same type of sample, leading to a lack of additional insight following plots using the same template. The results illustrated confirm a very close match between the results of the LS-ECM and DE-ECM method, which is also further supported by the convergence to highly similar parameter values between the two procedures summarized for sample 6 in Table 3. The same comparisons in Tables 2 and 3 as well as Figures 8 and 9 also indicate the LS-ARX procedure is substantially less accurate.

Table 2. Mean squared error (MSE) results.

Sample No.	SOC Range	Discharge Time (s)	LS-ARX	LS-ECM	DE-ECM (Average)
Sample 1	24.6–24.3%	10 s	5.349×10^{-7}	5.900×10^{-8}	5.890×10^{-8}
Sample 2	64.8–64.5%	10 s	4.164×10^{-7}	4.661×10^{-8}	4.660×10^{-8}
Sample 3	95.0–94.7%	10 s	5.971×10^{-7}	3.798×10^{-7}	3.794×10^{-7}
Sample 4	24.3–14.6%	~334 s	1.082×10^{-4}	4.825×10^{-6}	4.826×10^{-6}
Sample 5	64.5–54.8%	~334 s	2.120×10^{-5}	8.061×10^{-7}	8.058×10^{-7}
Sample 6	94.7–84.9%	~334 s	2.391×10^{-5}	7.069×10^{-6}	7.073×10^{-6}

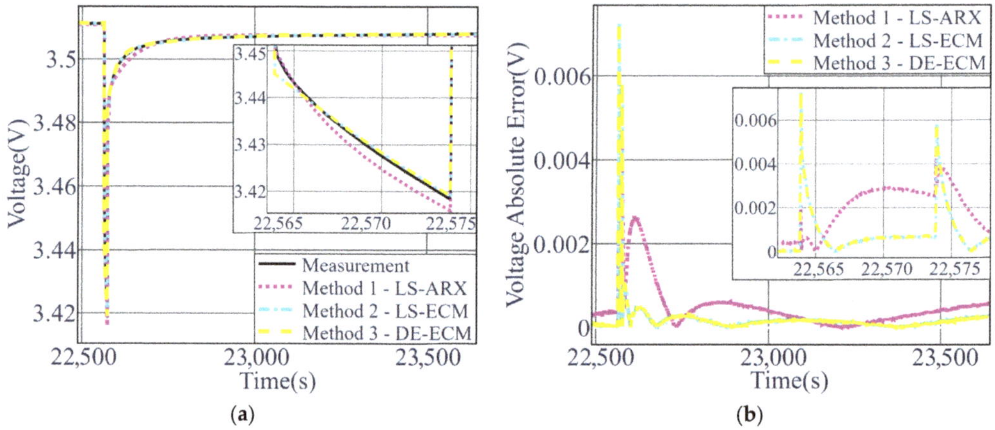

Figure 8. Sample 1 results. (**a**) Measured vs. simulated voltage; (**b**) voltage absolute error profile.

Table 3. Parameter values obtained for sample 6.

Parameter	LS-ARX	LS-ECM	DE-ECM (Average)
$V_{OC,max}$ (V)	4.103	4.108	4.108
$V_{OC,min}$ (V)	4.075	4.076	4.076
R_0 (Ω)	2.513×10^{-2}	2.666×10^{-2}	2.670×10^{-2}
R_1 (Ω)	1.011×10^{-2}	1.434×10^{-2}	1.437×10^{-2}
τ_1 (s)	3.850	13.788	13.938
R_2 (Ω)	2.240×10^{-2}	1.668×10^{-2}	1.664×10^{-2}
τ_2 (s)	95.961	183.044	184.345

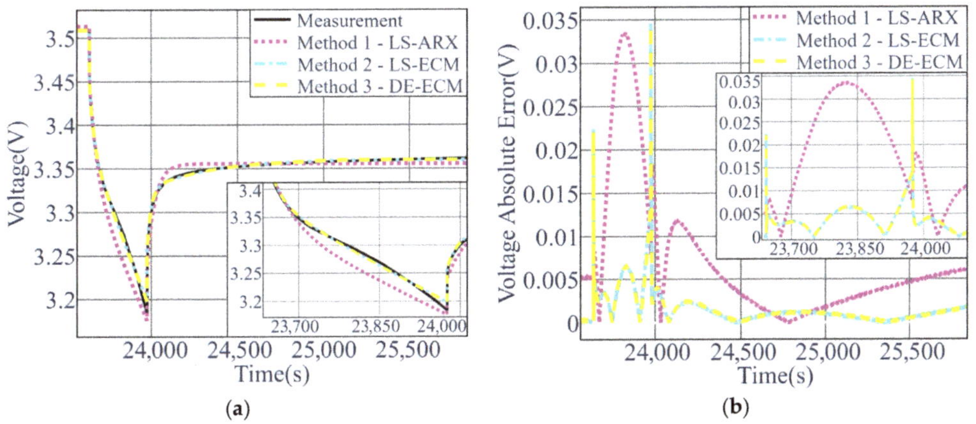

Figure 9. Sample 4 results. (**a**) Measured vs. simulated voltage; (**b**) voltage absolute error profile.

Targeting an efficiency perspective, the evolution of all three algorithms is also presented in Figure 10 for sample 1, while Table 4 summarizes the number of objective function evaluations for all samples. In addition to the absolute value, the number of evaluations was also expressed in percentages relative to the values indicated for the DE-ECM process. The results highlight the LS-ARX as the most effective from the perspective of computational expense due to its lowest MSE during the initial iterations. Nevertheless, the process also presents the lowest amount of improvement, leading to results with the highest final MSE. These attributes recommend the procedure to be more suited for value estimation tasks than highly accurate parameter extraction. This statement is also supported by the fact that the procedure requires no starting values while being able to compute all parameter values with one single computation.

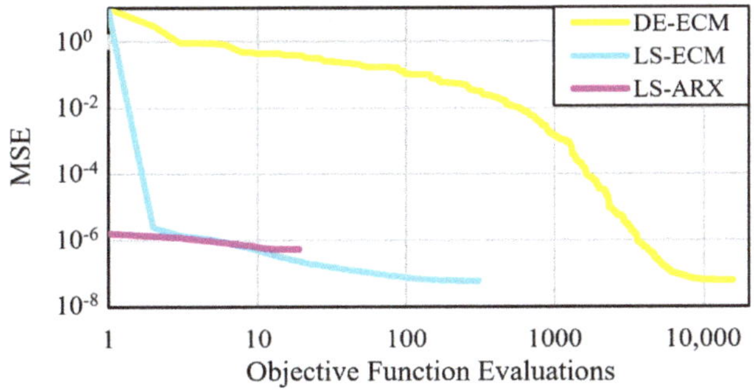

Figure 10. MSE evolution results (minimum MSE after each evaluation for sample 1).

Table 4. Number of objective function evaluations.

Sample No.	LS-ARX (% of DE-ECM)	LS-ECM (% of DE-ECM)	DE-ECM (Average)
Sample 1	19 (0.15%)	311 (2.42%)	12,836 (100%)
Sample 2	25 (0.17%)	161 (1.12%)	14,398 (100%)
Sample 3	20 (0.13%)	223 (1.48%)	15,107 (100%)
Sample 4	9 (0.08%)	121 (1.10%)	10,983 (100%)
Sample 5	7 (0.05%)	163 (1.15%)	14,129 (100%)
Sample 6	18 (0.16%)	77 (0.67%)	11,367 (100%)

The DE-ECM method can be observed as having the highest computational effort to converge to a satisfactory result. By considering the comparison study presented by Hou et al. [21], the evolution of the DE algorithm is representative of a global solver. Nevertheless, it should also be noted that substantial variations can be caused by the starting values and boundaries as well as the stochastic components in the optimization. The LS-ECM procedure presents the best compromise, with results directly comparable with global optimization, but with only a fraction of the computational cost, endorsing the procedure as highly effective in obtaining accurate results.

The results presented for the DE-ECM process in this section represent the average values obtained across the 10 runs computed for each sample. While the process has consistently converged to the same minimum point for all samples, the number of iterations required varied. Across all runs, the average percentage deviation was 7.942%, while the largest difference was found to be 25.678%, observed for one of the runs computed for sample 6. By contrast, the average percentage deviation for MSE, at 0.044%, was considered almost insignificant. For parameters, using sample 6 for exemplification, the values followed the sensitivity of the voltage response with respect to the type of parameter. As a result, the average for Open Circuit Voltage was found to be 8.173×10^{-4}%, for resistance values the average was 0.532%, while the highest average, at a value of 2.533%, was presented for time constants.

4. Discussion

4.1. Analysis of LS-ARX Procedure

Without any preprocessing in the cases presented, the initial LS-ARX results will be likely biased towards the high-frequency behavior found in the data. Applying a moving average filter has the benefit of not only removing noise from the data but also dampening the high-frequency dynamics captured, consequently making the slow dynamics increasingly more observable. In return, the slow time constants influence a higher percentage of the data and are more effective in reducing the MSE values driving the process. The downsampling process creates a similar transition, however, by adding simplification of the response signal. Accidental aliasing can also potentially benefit the same transition as the selection of the filter window size and the downsampling process is performed iteratively and not analytically. Despite improvement, the LS-ARX method still presents the least accurate results, as the issues identified in Section 2.2.1 are reduced by the optimization-type process, but not completely removed.

The results in Table 5 summarizing the values for filter window size and the magnitude of the downsampling (DS) process associated with the lowest error highlight that data segments containing short discharge steps (samples 1 to 3) benefit more from the optimization approach and reach the upper limit for the filter window size. By contrast, the long discharge pulses (samples 4 to 6) gain significantly less from the optimization approach, as only results from sample 6 present a significant improvement. One potential cause for this difference can be traced back to a reduced capability of the model to replicate the data caused by the long data segments containing more complex nonlinear behavior. In the case of LS applied to the ARX format, the margin between the model and data can also be considered to have a similar effect on error propagation. Nevertheless, as the preprocessing is intended to mostly attenuate high-frequency parts of the voltage signal, it will be ineffective in bringing an improvement for this separate cause. Attempting to find parameter values for two RC elements, which may present a significant gap in time scales using the same preprocessing setting, may also have a negative effect set to be reduced by a model decoupling-type approach.

Table 5. Global optimum preprocessing settings.

Sample No.	Default MSE	Best Result MSE	Filter Window Size	DS Factor	Global Minimum MSE
Sample 1	1.611×10^{-6}	5.349×10^{-7}	2048	2	5.316×10^{-7}
Sample 2	2.334×10^{-6}	4.164×10^{-7}	2048	16	4.043×10^{-7}
Sample 3	2.994×10^{-6}	5.971×10^{-7}	1024	4	5.971×10^{-7}
Sample 4	1.085×10^{-4}	1.082×10^{-4}	1	2	1.082×10^{-4}
Sample 5	2.120×10^{-5}	2.120×10^{-5}	1	1	1.705×10^{-5}
Sample 6	3.150×10^{-5}	2.391×10^{-5}	64	8	2.226×10^{-5}

To further investigate the limitations of the process, the results in Table 5 have also been enhanced with the global minimum achievable using the LS-ARX method. The complete mapping of the MSE error for sample 6 is presented in Figure 11. These bring to attention an additional limitation of the search algorithm employed, namely the very rudimentary optimization procedure of the preprocessing settings, which will likely converge to the local minima. Nevertheless, with the exemption of sample 5, the results are still comparable to the global MSE and require substantially fewer evaluations than evaluating all possible combinations.

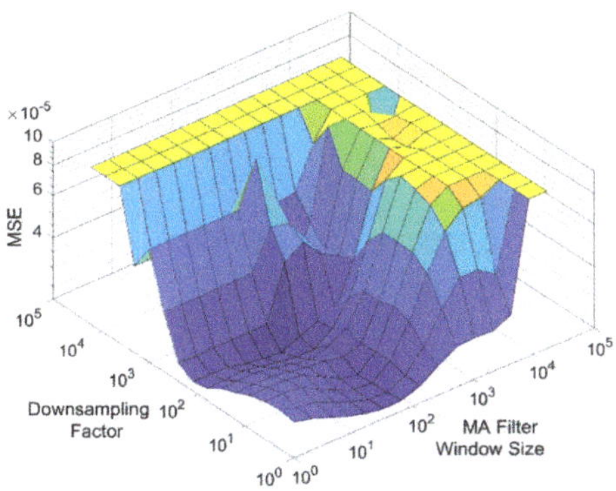

Figure 11. MSE function shape for sample 6—LS-ARX process (MSE limited at 10^{-5}).

The variation in MSE relative to the preprocessing settings illustrated in Figure 11 also suggests the procedures employed as potential limiting factors. The selection of moving average filtering and downsampling directly influence the shape of the MSE and may result in a shape with multiple local minima requiring to be found and evaluated. As a result, preprocessing methodologies such as frequency domain filters and decimation should also be considered as alternatives. Their efficiency should be evaluated in relation to LS-ARX using the achievable MSE as well as its shape relative to the parameters controlling them.

4.2. Analysis of LS-ECM Procedure

The second procedure introduced in this study endorses the application of the same LS process directly to the default state-space format of the model. The results present this approach as marginally less accurate in comparison to the one achieved by the DE-ECM procedure considered as a benchmark. The similarity of the results can be attributed to the shared elements between the two methodologies, such as the objective function and the iterative structure cascading information from one cycle to the next. The substantial gap in computation performance summarized by Table 4 also presents the main differentiator that caused the ability of the LS-ECM to include gradient information.

Nevertheless, the integration of a gradient-based approach for the second step of the LS-ECM means the capability of converging correctly is present only in the case of local optimization. The similarity between the results presented by the LS-ECM and DE-ECM for all six samples presents the first suggestion of correct convergence towards the global minima, and implicitly for the optimization problem to only present one local minima. For confirmation, additional examination of the procedure brought further evidence to support this hypothesis and prove the correct application of the process.

Figure 12 illustrates the shape of the objective function in the case of sample 6 during the last iteration, relative to the nonlinear parameters optimized in step 2 of the procedure. Step 1 parameters were not included as they are linear parameters, which will not reveal any risk of local minimum values being present. The shape of the objective function, despite presenting variability caused by the data segment, will consistently feature a main global minimum and a number of additional local minima dependent on the order of the RC model. In the case of the 2RC, only one additional local minima is present. For each set of linear parameters, a global minimum will be present, while a second will result from a potential inversion of the time constant values between them. As a result, the shape will present a degree of symmetry directly linked with the relation between the resistance values associated with the RC element (perfect symmetry will be caused by two equal values). It should be noted also that converging to the local minima instead of the global is not possible since each iteration of the first step in the process will set the resistance values to the global minimum for the current values of the time constants. This will lead to a global minimum occurring in the MSE surface created for step 2 in the proximity of their current values, rather than the mirrored part, supporting the optimization problem being correctly classified as local. It should be noted that an additional requirement for the process to function in the described manner is for the data points to be equidistant in the time axis, as the presence of data spaced unequally will introduce additional local minima points. Such unequal spacing can result from data compression algorithms and should be corrected as a preprocessing step.

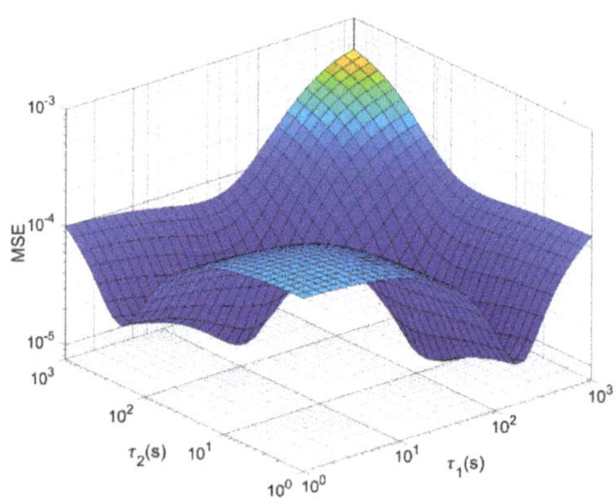

Figure 12. MSE function shape for sample 6—LS-ECM process (step 2; cycle 38).

The surface illustrated in Figure 12 also highlights a separate limitation, namely the convergence, which is naturally dependent on the gradient of the objective function. This will affect convergence speed and overall robustness since small gradients will reduce significantly the efficiency of the process. In the extreme, gradients that approach zero present the risk of instability caused by insufficient precision in the computation. The potential issue created by gradient should also be accounted for during initialization; in the current study, the lower bound of 0 for the time constant values encouraged initial values relatively close to this value. This increased the likelihood of undervalued time constant parameters effective in lowering MSE quickly as they increase, meaning a sufficiently observable gradient for the computations to be stable.

5. Conclusions

Two new methodologies for the identification of ECM parameters from battery systems data based on least squares linear regression (LS) have been proposed, analyzed and compared against a third indicated by prior research as a benchmark and represented by global optimization, namely differential evolution. Both methods presented satisfactory results as well as a trade-off between the effort and knowledge required to set up, computational effort, and accuracy of results. The best compromise is represented by the LS-ECM process employing the application of the linear solver directly to the state-space discrete equations of the model. The process is enabled by effective linearization for the nonlinear parameters through recursive computation of the total derivative. This procedure presented limited computational effort and knowledge to set up while producing nearly identical results with respect to the global optimization in terms of accuracy. The highest MSE of 3.798×10^{-7} V^2 and 7.069×10^{-6} V^2 was obtained over short and long pulses of the current part of cell-level data obtained from a Sony VTC6 cell. The match between the results of the process and the results of the global optimization also suggests that the error is caused by the capability of the model to replicate the system (linear response assumption) rather than the parameter identification method employed.

By comparison, the LS-ARX process presented as the fastest, while also extracting parameter values directly from data, without any assumption over the starting values. Nevertheless, the limitations presented by the recursive equations of the model result in penalties in accuracy, making this method recommendable for fast extraction of parameter estimates to be further refined. The DE-ECM process is the most robust and accurate solution, as well as the easiest to set up, but also very computationally expensive, potentially prohibiting the use of the procedure in the case of large datasets.

Future work will target the implementation of the extraction methods to build a complete battery system model, and is set to include the nonlinear behavior associated with temperature, SOC, and current amplitude. Moreover, a more efficient extraction procedure can balance the benefits of the proposed methods by setting the starting point for the LS-ECM by using the LS-ARX. A more comprehensive understanding of the reduction in computational effort may be provided by adding to the comparison a state-of-the-art local optimization, while each procedure also represents the base for further enhancements. In the case of the LS-ARX process, alternative optimization and preprocessing procedures may present interest. For LS-ECM, hybridization with a different optimization structure may focus on accelerated convergence, reducing the gradient-induced issues in relation to starting points and convergence speed. Lastly, the LS-ECM procedure may also present interest as the fundamental element supporting an alternative RLS-type procedure.

Author Contributions: Methodology: V.-I.S., G.E., and P.W.; software: V.-I.S.; validation: G.E., C.B., and N.D.; resources: C.B., P.W., E.K., and N.Z.; writing—draft preparation: V.-I.S.; writing—review and editing: G.E., N.D., C.B., and P.W.; supervision C.B., N.Z., E.K., and P.W. All authors have read and agreed to the published version of the manuscript.

Funding: This research was funded by the EPSRC Centre for Doctoral Training in Advanced Automotive Propulsion Systems (AAPS), under project EP/S023364/1 and AVL List GmbH.

Data Availability Statement: The software presented in this article is not readily available following restrictions imposed by commercialization. The raw cell data supporting the conclusions of this article are available on request from the corresponding author.

Conflicts of Interest: Authors Georg Engel, Nico Didcock and Emre Kural were employed by the company AVL List GmbH. The remaining authors declare that the research was conducted in the absence of any commercial or financial relationships that could be construed as a potential conflict of interest. The authors declare that this study received funding from AVL List GmbH. The funder was not involved in the study design, collection, analysis, interpretation of data, the writing of this article or the decision to submit it for publication.

Appendix A

To assist the model equation formulation, the following have to be summarized:

- Kirchhoff's First Law applied to an RC element (Equation (A1))

$$I_i(t) + I_{C_i}(t) - I_t(t) = 0 \tag{A1}$$

- Kirchhoff's Second Law applied to an RC element (Equation (A2))

$$V_i(t) - V_{C_i}(t) = 0 \tag{A2}$$

- Capacitive current definition (Equation (A3))

$$I_{C_i}(t) = C_i \frac{dV_{C_i}(t)}{dt} \tag{A3}$$

Equation (A4) captures the derivative of both terms in Equation (A2) and replaces the $\frac{dV_{C_i}(t)}{dt}$ term using Equation (A3). The next step substitutes the $I_{C_i}(t)$ according to Equation (A1), resulting in Equation (A5). The last step, resulting in Equation (A6) or Equation (2) in the main text, is presented by the application of Ohm's law to Equation (A4) and applying the $\tau_i = R_i C_i$ substitution.

$$\frac{dV_i(t)}{dt} = \frac{dV_{C_i}(t)}{dt} = \frac{I_{C_i}(t)}{C_i} \tag{A4}$$

$$\frac{dV_i(t)}{dt} \times \frac{1}{R_i} = \left(\frac{I_t(t) - I_i(t)}{R_i C_i} \right) \tag{A5}$$

$$\frac{dI_i(t)}{dt} = \frac{-1}{\tau_i} I_i(t) + \frac{1}{\tau_i} I_t(t) \tag{A6}$$

Appendix B

The discretization of Equation (2) of the model has been achieved using the simplistic forward Euler's method, detailed by Zeigler, B. [25] and summarized in Equation (A7). The application of the process is captured by Equation (A8), while Equation (A9) summarizes the final equation, defined as Equation (3) in the main text.

$$f[t, y(t)] = \frac{dy}{dt} \cong \frac{y(t + \Delta t) - y(t)}{\Delta t} \tag{A7}$$

$$\frac{-1}{\tau_i} I_i(t_k) + \frac{1}{\tau_i} I_0(t_k) \cong \frac{I_i(t_{k+1}) - I_i(t_k)}{\Delta t} \tag{A8}$$

$$1 - \frac{\Delta t}{\tau_i} I_i(t_k) + \frac{\Delta t}{\tau_i} I_0(t_k) \cong I_i(t_{k+1}) \tag{A9}$$

Appendix C

The conversion to an ARX format for the discretized version of the model starts by defining a sub-model of the original capturing the voltage drop associated with the input current in equation (A10). This only removes the V_{OC} term of Equation (1), meaning it can be easily rearranged in the state-space format defined in Equations (A11) and (A12), following the substitutions indicated below.

$$V_B(t_k) = \sum_{i=1}^{n} R_i I_i(t_k) + R_0 I_0(t_k) \tag{A10}$$

$$x[k+1] = Ax[k] + Bu[k] \tag{A11}$$

$$y[k] = Cx[k] + Du[k] \tag{A12}$$

$$A = \begin{bmatrix} 1-\frac{\Delta t}{\tau_1} & 0 & \cdots & 0 \\ 0 & 1-\frac{\Delta t}{\tau_2} & & 0 \\ \vdots & & \ddots & \vdots \\ 0 & 0 & \cdots & 1-\frac{\Delta t}{\tau_n} \end{bmatrix}; B = \begin{bmatrix} \frac{\Delta t}{\tau_1} \\ \frac{\Delta t}{\tau_2} \\ \vdots \\ \frac{\Delta t}{\tau_n} \end{bmatrix}; C = \begin{bmatrix} R_1 & 0 & \cdots & 0 \\ 0 & R_2 & & 0 \\ \vdots & & \ddots & \vdots \\ 0 & 0 & \cdots & R_n \end{bmatrix}; D = R_0$$

$$x[k] = \begin{bmatrix} I_1(t_k) \\ I_2(t_k) \\ \vdots \\ I_n(t_k) \end{bmatrix}; \; u[k] = I_0(t_k); \; y[k] = V_B(t_k);$$

The following application of the z-transform is detailed by Cheever, E. [26] while considering no response to the initial conditions (set to 0), resulting in Equations (A13) and (A14).

$$zX[z] = AX[z] + BU[z] \tag{A13}$$

$$Y[z] = CX[z] + DU[z] \tag{A14}$$

Equations (A15)–(A17) present the rearrangement of terms in Equations (A13) and (A14) to support the definition of the discrete transfer function $G[z]$ in Equation (A18), in line with the instructions supplied by Cheever, E. [27]

$$(zI - A)X[z] = BU[z] \tag{A15}$$

$$X[z] = (zI - A)^{-1}BU[z] \tag{A16}$$

$$Y[z] = \left[C(zI - A)^{-1}B + D\right]U[z] \tag{A17}$$

$$G[z] = \frac{Y[z]}{U[z]} = \left[C(zI - A)^{-1}B + D\right] = \frac{\sum_{i=0}^{n} b_i\prime z^{-i}}{\sum_{i=0}^{n} a_i\prime z^{-i}} \tag{A18}$$

Cross-multiplication of the fractions in Equation (A18) will present the ARX model captured in Equation (A19). This is presented following the return of the substitutions used in support of Equations (A11) and (A12) and adopting the substitution $a_i = \frac{a_i'}{a_0'}$ for the $a_i\prime$ and $b_i\prime$ to remove the coefficient of the $Y[z]$ replaced by $V_B(z)$.

$$V_B(z) = \sum_{i=0}^{n} b_i I_0 z^{-i} - \sum_{i=1}^{n} a_i V_B z^{-i} \tag{A19}$$

Starting from Equation (A19), the model can be manipulated to contain the constant $V_{OC}(z)$ term following Equations (A20) and (A21). The substitution captured by Equation (A22) results in Equation (4) in the main text.

$$V_{OC}(z) - V_B(z) = V_{OC}(z) - \sum_{i=0}^{n} b_i I_0 z^{-i} + \left(\sum_{i=1}^{n} a_i V_{OC} z^{-i} - \sum_{i=1}^{n} a_i V_{Out} z^{-i}\right) \tag{A20}$$

$$V_{OC}(z) - V_B(z) = V_{OC}\left(1 + \sum_{i=1}^{n} a_i\right) - \sum_{i=1}^{n} a_i V_{Out} z^{-i} - \sum_{i=0}^{n} b_i I_0 z^{-i} \tag{A21}$$

$$c = V_{OC}(z)(1 + \sum_{i=1}^{n} a_i) \tag{A22}$$

References

1. Kalogiannis, T.; Hosen, M.S.; Sokkeh, M.A.; Goutam, S.; Jaguemont, J.; Jin, L.; Qiao, G.; Berecibar, M.; Van Mierlo, J. Comparative Study on Parameter Identification Methods for Dual-Polarization Lithium-Ion Equivalent Circuit Model. *Energies* **2019**, *12*, 4031. [CrossRef]
2. Li, J.; Peng, Y.; Wang, Q.; Liu, H. Status and Prospects of Research on Lithium-Ion Battery Parameter Identification. *Batteries* **2024**, *10*, 194. [CrossRef]
3. Ljung, L. *System Identification*; Webster, J.G., Ed.; Wiley Encyclopedia of Electrical and Electronics Engineering: Hoboken, NJ, USA, 1999. [CrossRef]
4. Zhang, T.; Guo, N.; Sun, X.; Fan, J.; Yang, N.; Song, J.; Zou, Y. A Systematic Framework for State of Charge, State of Health and State of Power Co-Estimation of Lithium-Ion Battery in Electric Vehicles. *Sustainability* **2021**, *13*, 5166. [CrossRef]
5. Huang, C.-S. Online Parameter Identification for Lithium-Ion Batteries: An Adaptive Moving Window Size Design Methodology for Least Square Fitting. *IEEE Trans. Veh. Technol.* **2023**, *72*, 5824. [CrossRef]
6. Raihan, S.A.; Balasingam, B. Recursive Least Square Estimation Approach to Real-Time Parameter Identification in Li-ion Batteries. In Proceedings of the 2019 IEEE Electrical Power and Energy Conference (EPEC), Montreal, QC, Canada, 16–18 October 2019. [CrossRef]
7. Kim, M.; Kim, K.; Han, S. Reliable Online Parameter Identification of Li-Ion Batteries in Battery Management Systems Using the Condition Number of the Error Covariance Matrix. *IEEE Access* **2020**, *8*, 189106. [CrossRef]
8. Fan, Y.; Shi, H.; Wang, S.; Fernandez, C.; Cao, W.; Huang, J. A Novel Adaptive Function—Dual Kalman Filtering Strategy for Online Battery Model Parameters and State of Charge Co-Estimation. *Energies* **2021**, *14*, 2268. [CrossRef]
9. Xia, B.; Huang, R.; Lao, Z.; Zhang, R.; Lai, Y.; Zheng, W.; Wang, H.; Wang, W.; Wang, M. Online Parameter Identification of Lithium-Ion Batteries Using a Novel Multiple Forgetting Factor Recursive Least Square Algorithm. *Energies* **2018**, *11*, 3180. [CrossRef]
10. Sun, X.; Ji, J.; Ren, B.; Xie, C.; Yan, D. Adaptive Forgetting Factor Recursive Least Square Algorithm for Online Identification of Equivalent Circuit Model Parameters of a Lithium-Ion Battery. *Energies* **2019**, *12*, 2242. [CrossRef]
11. Sun, C.; Lin, H.; Cai, H.; Gao, M.; Zhu, C.; He, Z. Improved parameter identification and state-of-charge estimation for lithium-ion battery with fixed memory recursive least squares and sigma-point Kalman filter. *Electrochim. Acta* **2021**, *387*, 138501. [CrossRef]
12. Li, C.; Kim, G.-W. Improved State-of-Charge Estimation of Lithium-Ion Battery for Electric Vehicles Using Parameter Estimation and Multi-Innovation Adaptive Robust Unscented Kalman Filter. *Energies* **2024**, *17*, 272. [CrossRef]
13. Xu, Y.; Chen, X.; Zhang, H.; Yang, F.; Tong, L.; Yang, Y.; Yan, D.; Yang, A.; Yu, M.; Liu, Z.; et al. Online identification of battery model parameters and joint state of charge and state of health estimation using dual particle filter algorithms. *Int. J. Energy Res.* **2022**, *46*, 19615–19652. [CrossRef]
14. Tian, N.; Wang, Y.; Chen, J.; Fang, H. On parameter identification of an equivalent circuit model for lithium-ion batteries. In Proceedings of the IEEE Conference on Control Technology and Applications (CCTA), Maui, HI, USA, 27–30 August 2017. [CrossRef]
15. Tran, M.-K.; DaCosta, A.; Mevawalla, A.; Panchal, S.; Fowler, M. Comparative Study of Equivalent Circuit Models Performance in Four Common Lithium-Ion Batteries: LFP, NMC, LMO, NCA. *Batteries* **2021**, *7*, 51. [CrossRef]
16. Chen, C. *Parameter Determination for the Battery Equivalent Circuit Model Using a Numerical Integro-Differential Method*; SAE Technical Paper 2020-01-1179; SAE International: Warrendale, PA, USA, 2020. [CrossRef]
17. Al Rafei, T.; Yousfi Steiner, N.; Chrenko, D. Genetic Algorithm and Taguchi Method: An Approach for Better Li-Ion Cell Model Parameter Identification. *Batteries* **2023**, *9*, 72. [CrossRef]
18. Zhou, S.; Liu, X.; Hua, Y.; Zhou, X.; Yang, S. Adaptive model parameter identification for lithium-ion batteries based on improved coupling hybrid adaptive particle swarm optimization- simulated annealing method. *J. Power Sources* **2021**, *482*, 228951. [CrossRef]
19. Ghoulam, Y.; Mesbahi, T.; Wilson, P.; Durand, S.; Lewis, A.; Lallement, C.; Vagg, C. Lithium-Ion Battery Parameter Identification for Hybrid and Electric Vehicles Using Drive Cycle Data. *Energies* **2022**, *15*, 4005. [CrossRef]
20. Cheng, Y.S. Identification of Parameters for Equivalent Circuit Model of Li-Ion Battery Cell with Population Based Optimization Algorithms. Available online: https://ssrn.com/abstract=4229575 (accessed on 14 September 2024).
21. Hou, J.; Wang, X.; Su, Y.; Yang, Y.; Gao, T. Parameter Identification of Lithium Battery Model Based on Chaotic Quantum Sparrow Search Algorithm. *Appl. Sci.* **2022**, *12*, 7332. [CrossRef]
22. Lee, S.; Lee, D. A Novel Battery State of Charge Estimation Based on Voltage Relaxation Curve. *Batteries* **2023**, *9*, 517. [CrossRef]
23. Lee, J.; Sun, H.; Liu, Y.; Li, X. A machine learning framework for remaining useful lifetime prediction of li-ion batteries using diverse neural networks. *Energy AI* **2024**, *15*, 100319. [CrossRef]
24. Cleary, T.; Nozarijouybari, Z.; Wang, D.; Wang, D.; Rahn, C.; Fathy, H.K. An Experimentally Parameterized Equivalent Circuit Model of a Solid-State Lithium-Sulfur Battery. *Batteries* **2022**, *8*, 269. [CrossRef]
25. Zeigler, B.; Muzy, A.; Kofman, E. Chapter 3—Modeling Formalisms and Their Simulators. In *Theory of Modeling and Simulation*, 3rd ed.; Zeigler, B., Muzy, A., Kofman, E., Eds.; Academic Press: Cambridge, MA, USA, 2019; pp. 43–91.
26. Cheever, E. The Z Transform. 2022. Available online: https://lpsa.swarthmore.edu/ZXform/FwdZXform/FwdZXform.html (accessed on 14 September 2024).
27. Cheever, E. Transformation: Transfer Function ↔ State Space. 2022. Available online: https://lpsa.swarthmore.edu/Representations/SysRepTransformations/TF2SS.html (accessed on 14 September 2024).

28. SciPy. Scipy.optimize.lsq_linear—SciPy v1.10.1 Manual. 2023. Available online: https://docs.scipy.org/doc/scipy/reference/generated/scipy.optimize.lsq_linear.html (accessed on 14 September 2024).
29. SciPy. Scipy.optimize.differential_evolution—SciPy v1.10.1 Manual. 2023. Available online: https://docs.scipy.org/doc/scipy/reference/generated/scipy.optimize.differential_evolution.html (accessed on 14 September 2024).
30. PyVisa Authors. PyVISA—Control Your Instruments with Python, PyVISA. 2023. Available online: https://pyvisa.readthedocs.io/en/latest/index.html (accessed on 14 September 2024).

batteries

MDPI

Article

Modeling Thermal Runaway Mechanisms and Pressure Dynamics in Prismatic Lithium-Ion Batteries

Mohammad Ayayda *, Ralf Benger, Timo Reichrath, Kshitij Kasturia, Jacob Klink and Ines Hauer

Research Center for Energy Storage Technologies, Clausthal University of Technology, 38678 Clausthal-Zellerfeld, Germany; ralf.benger@tu-clausthal.de (R.B.); timo.reichrath@tu-clausthal.de (T.R.); ines.hauer@tu-clausthal.de (I.H.)

* Correspondence: mohammad.ayayda@tu-clausthal.de

Abstract: Lithium-ion batteries play a vital role in modern energy storage systems, being widely utilized in devices such as mobile phones, electric vehicles, and stationary energy units. One of the critical challenges with their use is the thermal runaway (TR), typically characterized by a sharp increase in internal pressure. A thorough understanding and accurate prediction of this behavior are crucial for improving the safety and reliability of these batteries. To achieve this, two new combined models were developed: one to simulate the thermal runaway and another to simulate the internal cell pressure. The thermal model tracks a chain of decomposition reactions that eventually lead to TR. At the same time, the pressure model simulates the proportional increase in pressure due to the evaporation of the electrolyte and the gases produced from the decomposition reactions. What sets this work apart is the validation of the pressure model through experimental data, specifically for prismatic lithium-ion cells using NMC chemistries with varying stoichiometries—NMC111 and NMC811. While the majority of the literature focuses on the simulation of temperature and pressure for cylindrical cells, studies addressing these aspects in prismatic cells are much less common. This article addresses this gap by conducting pressure validation experiments, which are hardly documented in the existing studies. Furthermore, the model's accuracy and flexibility are tested through two experiments, conducted under diverse conditions to ensure robust and adaptive predictions of cell behavior during failure scenarios.

Keywords: lithium-ion battery; thermal runaway; internal pressure; simulation; modeling

1. Introduction

In recent years, the rise of electric vehicles has further intensified the demand for powerful and reliable energy storage systems. Lithium-ion batteries, renowned for their high energy density and efficiency, have become the cornerstone of this e-mobility revolution. Their ability to deliver substantial power with minimal weight, coupled with low self-discharge and negligible memory effects, makes them ideal for the rigorous demands of EVs. As a result, these batteries are now integral to the evolution of sustainable transportation [1]. Simultaneously, the global transition to renewable energy sources has underscored the need for efficient energy storage solutions. Lithium-ion technology stands out as the preferred option due to its cost-effectiveness and reliability, supporting a wide range of applications across industries. This dual role in both e-mobility and renewable energy integration illustrates the versatility and importance of lithium-ion batteries in today's energy landscape [2]. However, despite these benefits, Li-ion batteries also pose significant safety risks, particularly related to the phenomenon of thermal runaway (TR). TR can be triggered by various factors, such as thermal disturbances, overcharging, mechanical damage, or internal short circuits [3]. This phenomenon leads to uncontrolled temperature increases and a rapid rise in internal pressure due to the formation of gases from decomposition reactions and electrolyte evaporation [4–6]. If not promptly controlled in an adequate manner, these processes can result in severe consequences, including fires or explosions [7]. Therefore, understanding the mechanism of thermal runaway is crucial for developing models that can replicate TR processes and predict associated pressure developments within the cell during a fault. By predicting thermal runaway, solutions can be developed to avoid its occurrence. For instance, calculating the cooling power required to prevent TR can serve as a basis for developing more effective thermal management systems. Thermal runaway consists of a sequence of reactions leading to the cell's self-heating and culminating in TR [8]. When the cell is triggered and its temperature exceeds 50 °C, the decomposition of the solid electrolyte interphase (SEI) initiates, producing gases such as C_2H_4, CO_2, O_2 and H_2 [9,10]. Further temperature increases expose the anode to the electrolyte, causing the formation of a new SEI layer, a process known as SEI regeneration. This process occurs in a temperature range of 120–250 °C and is associated with additional gas production, including C_3H_6, C_2H_2, and CH_4 [11,12]. The cell becomes unstable following the complete collapse of the separator, which leads to an internal short circuit. The melting point of the separator depends on the materials used. For instance, a multi-layer separator made from polypropylene (PP) and polyethylene (PE) will start to melt at around 127 °C for the PE layer and at approximately 165 °C for the PP layer [13–15]. Subsequent temperature increases lead to further decomposition of the electrolyte and cathode material, generating significant amounts of gases such as CO, O_2, and HF [9,10].

Correspondingly, the generation of gases and the evaporation of the electrolyte cause a rise in internal pressure within the cell. Figure 1 underscores the components responsible for pressure changes both before and after the venting event. Initially, the rise in pressure is due to the evaporation of the electrolyte, occurring prior to the onset of decomposition reactions. Subsequently, the pressure continues to increase as these decomposition reactions commence and more electrolyte evaporates [16]. When the pressure surpasses a critical threshold, the rupture disc bursts, opening the cell and releasing the gases, leading to first venting [5]. This gas outflow removes part of the heat from the cell, causing a temperature drop. The internal pressure falls to ambient levels but rises again rapidly due to gas generation from TR-driven decomposition reactions, as shown in Figure 1 and confirmed by Zhang et al. [17] and Mao et al. [18].

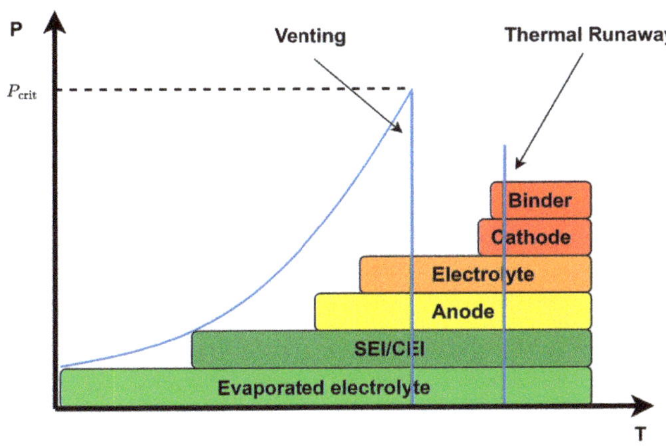

Figure 1. Illustration of the pressure increase in the cell and the contributing components. The diagram highlights the critical pressure point P_{crit} at which the cell opens and venting takes place, highlighting the elements responsible for pressure increases both prior to and following this event. Illustration style and insights are adapted from [19,20].

The current literature on the simulation and modeling of temperature and pressure during thermal runaway in prismatic cells is relatively sparse, with a predominant focus on cylindrical cells [21], as illustrated in Table 1. However, an exception to this trend is the study conducted by Wang et al. [22], which examined a 27 Ah Lithium iron phosphate (LFP) prismatic cell subjected to thermal abuse. In their study, the cell was heated from the side with a 2000 W electric heater until thermal runaway was induced. The model developed by Wang et al. primarily addressed the simulation of temperature distributions, jet velocity, and mass loss during thermal abuse conditions. Another significant study is by Feng et al. [23], which comprehensively discussed the electrochemical-thermal behavior of cells under thermal abuse and successfully developed a model to simulate temperature and voltage profiles during TR. Their work was validated against experimental results from accelerating rate calorimeter (ARC) tests on a 25 Ah Lithium nickel manganese cobalt oxide (NMC) cell. However, none of these studies measured or described in detail the development of internal pressure in prismatic cells during TR.

Table 1. Overview of thermal and pressure modeling approaches for various cell chemistries and geometries found in the literature.

Cathode Chemistry	Geometry	Modeling Approach	Pressure Validation	Ref.
NMC	Cylindrical	Lumped	Yes	[21]
NMC	Cylindrical	Lumped	Yes	[18]
NMC	Cylindrical	Lumped	No	[24]
NMC	Cylindrical	3D	No	[17]
NCA	Cylindrical	Lumped	No	[25]
LFP	Prismatic	Lumped	No	[22]
NMC	Pouch	Lumped	No	[16]

In contrast to the current literature, the validation of the model results presented in this paper was conducted using data obtained from two distinct experimental setups that involved prismatic cells from the same manufacturer. These experiments utilized NMC111 and NMC811 cells, each with diverse capacities and subjected to a range of test conditions, including different states of charge (SoC), heating rates, heating orientations, initial temperatures, and other variables. By incorporating these diverse experimental conditions, the aim was to systematically validate the model's accuracy and reliability. This comprehensive approach was undertaken to demonstrate the model's flexibility and to ensure that it performs consistently and reliably under different operational scenarios, ultimately confirming its robustness and adaptability to a variety of conditions encountered in practical applications.

The primary contributions of this work include the development and integration of a novel combination of thermal and pressure models derived from the existing literature. Additionally, the use of NMC prismatic cells, which are commonly employed in the automotive industry, provided a relevant context for validating the simulation results. The study also highlights the versatility and reliability of the model by examining its performance under various experimental conditions and with different cathode stoichiometries. A significant focus of the paper is the validation of pressure data using experimental measurements, addressing a gap in the current literature, as shown in Table 1.

The remainder of this paper is structured as follows: Section 2 details the development strategy of the thermal and pressure models. This is followed by a description of the experiments conducted to validate the simulation results, presented in Section 3. Subsequently, Section 4 discusses the comparison between the model outputs and experimental data. Finally, the paper concludes with a summary in Section 5.

2. Model Development

The model is composed of interconnected thermal and pressure submodels, designed to simulate cell behavior during a thermal runaway event. The thermal submodel quantifies heat generation from decomposition reactions and potential short circuits, while the pressure submodel simulates internal pressure increases due to electrolyte evaporation and decomposition gas production as temperature rises.

The model is constructed as lumped, non-adiabatic frameworks, focusing on a single cell during thermal abuse conditions. Within this framework, the specific heat capacity and convection coefficient are assumed constant throughout the simulations. Additionally, the masses of the various cell components are estimated using measurements from similar cells reported in the literature.

2.1. Thermal Model

The developed thermal model, illustrated in Figure 2, comprises several interconnected submodules that collaboratively simulate the temperature profile of the cell. A key feature of the model is its adaptable external heating submodule, which can be configured to meet varying experimental requirements. This flexibility allows the heating to be specified either in terms of power (W) or heating rate (K/min).

The initiation of self-heating within the cell is represented in the decomposition submodule. This submodule simulates the decomposition reaction chain analogous to that observed in the actual cell. Once the separator fails, leading to an internal short circuit, the model simulates the heat generated from both the redox reaction between the anode and cathode and the heat produced from the internal short circuit. The quantification of the heat from the internal short circuit is based on established frameworks provided by previous studies, namely, the work of Feng et al. [26], Hong et al. [27], and Ren et al. [28].

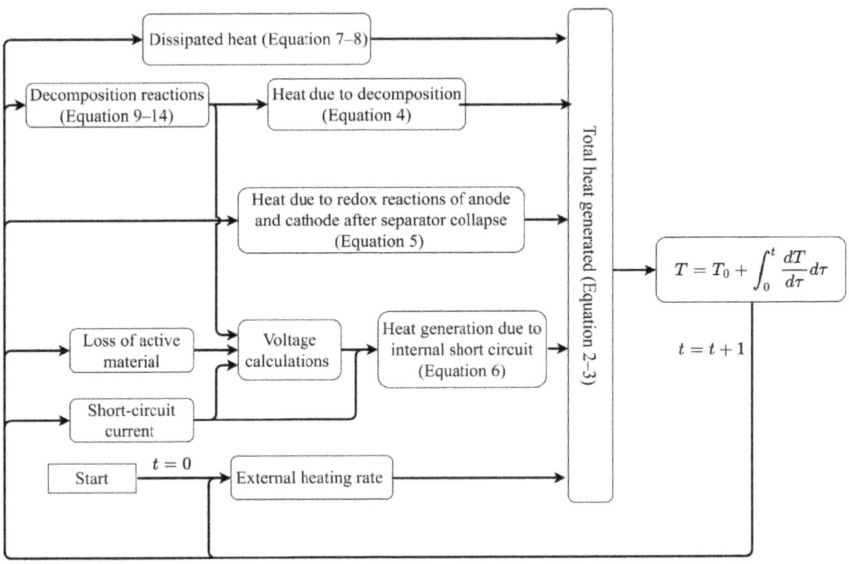

Figure 2. Flow chart of the thermal model considering heat generation within the cell, an external heat source, and heat dissipation (based on [23,26–28]).

2.1.1. Heat Generation in the Battery

The temperature T of the cell during TR can be calculated using [29]

$$T = T_0 + \int_0^t \frac{dT}{d\tau} d\tau \quad (1)$$

where T_0 is the initial temperature of the cell, and $\frac{dT}{dt}$ represents the rate of temperature change.

The rate of temperature change, $\frac{dT}{dt}$, can be determined using the following equation [23,30]:

$$\frac{dT}{dt} = \frac{\dot{Q}}{m_{\text{cell}} c_{\text{p}}} \tag{2}$$

where $\dot{Q}$ is the total heat flow rate generated through various chemical reactions; m_{cell} is the total mass of the cell; and c_p is the specific heat capacity.

$\dot{Q}$ is derived from several key sources, as shown in Equation (3):

$$\dot{Q} = \sum_{\text{z}} \dot{Q}_{\text{z}} + \dot{Q}_{\text{cat\&an}} + \dot{Q}_{\text{ISC}} - \dot{Q}_{\text{diss}} \tag{3}$$

where $\dot{Q}_{\text{z}}$ represents the heat flow rate resulting from the decomposition reactions of the various components inside the cell: SEI, anode, electrolyte, and cathode; $\dot{Q}_{\text{cat \& an}}$ is the additional heat flow rate generated due to the rapid oxidation-reduction reaction after separator collapse, leading to direct contact between the anode and cathode. This contact prompts a complex and rapid oxidation-reduction reaction, attributed to "chemical crosstalk", as proposed by Liu et al. [31] and Feng et al. [26]; $\dot{Q}_{\text{ISC}}$ denotes the heat flow rate released due to short-circuit conditions; $\dot{Q}_{\text{diss}}$ is the dissipated heat power into the environment. $\dot{Q}_{\text{z}}$ is dependent on the decomposition rate of the reactants, $\frac{dc_{\text{z}}}{dt}$, as shown in Equation (4).

$$\dot{Q}_{\text{z}} = \Delta h_{\text{z}} \cdot m_{\text{z}} \cdot \frac{dc_{\text{z}}}{dt} \tag{4}$$

Δh_{z} represents the specific enthalpy of the reactants; and m_{z} denotes the mass of the components participating in the reaction. The mass utilized in calculating the heat generated from SEI decomposition is equivalent to the mass of the anode. Similarly, the mass employed in determining the heat generation from the cathode electrolyte interphase (CEI) decomposition corresponds to the mass of the cathode [21,24,26].

The heat released due to short-circuit conditions is determined by [26–28]

$$\dot{Q}_{\text{cat\&an}} = \frac{1}{\Delta t}\left(\Delta H_{\text{cat\&an}} - \int_0^t \dot{Q}_{\text{cat\&an}}\, d\tau\right) \tag{5}$$

$$\dot{Q}_{\text{ISC}} = \varsigma \left(\frac{V}{R_{\text{ISC}}}\right)^2 (R_{\text{cell}} + R_{\text{ISC}}) \tag{6}$$

Δt is the average short circuit time; $\Delta H_{\text{cat\&an}}$ is the total heat generation from oxidation-reduction reactions; ς is an efficiency factor used to correct for non-ohmic heat generation, as proposed by Coman et al. [32]; V is the voltage; R_{ISC} is the equivalent resistance in the resistor–capacitor circuit model; and R_{cell} is the internal resistance of the cell. Detailed calculations of voltage and resistance are based on the work of Feng et al. [26].

The heat dissipated to the environment due to convection $\dot{Q}_{\text{conv}}$ and radiation $\dot{Q}_{\text{rad}}$ is calculated using the following equations [33,34]:

$$\dot{Q}_{\text{conv}} = -h_{\text{conv}} A_{\text{cell}} (T - T_{\text{a}}) \tag{7}$$

$$\dot{Q}_{\text{rad}} = -\varepsilon \sigma A_{\text{cell}} (T^4 - T_{\text{a}}^4) \tag{8}$$

Here , h_{conv} is the convection heat transfer coefficient; A_{cell} is the surface area of the cell; ε is the surface emissivity; σ is the Stefan–Boltzmann constant; and T_{a} is the ambient temperature.

Table 2 presents the values of the parameters and constants utilized in the heat generation submodel.

Table 2. Parameters of the heat generation submodel for the tested prismatic cells [11,23,26,35].

Parameter	Symbol	Value	Unit
Specific heat capacity	c_p	1100	$J\,kg^{-1}\,K^{-1}$
Total mass of the NMC111 cell	m_{NMC111}	810	g
Total mass of the NMC811 cell	m_{NMC811}	900	g
Time constant	Δt	10	s
Total heat generation from redox reactions	$\Delta H_{cat\&an}$	308	kJ
Efficiency factor	ς	0.28	-
Heat transfer coefficient	h_{conv}	5	$W\,m^{-2}\,K^{-1}$
Surface area of the battery	A_{cell}	135	cm^2
Emissivity	ε	0.8	-
Stefan-Boltzmann constant	σ	5.67×10^{-8}	$W\,m^{-2}\,K^{-4}$

2.1.2. Reactions Kinetics

The reaction rate of the exothermal reactions involving the cell components can be described using the Arrhenius equation [32,36]:

$$\frac{dc_z}{dt} = A_z \cdot c_z^{n_{z,1}} \cdot (1 - c_z)^{n_{z,2}} \cdot \exp\left(-\frac{E_{a,z}}{R_0 T}\right) \tag{9}$$

where A_z is the frequency factor; c_z is the normalized concentration of the reactants; n_z is the reaction order; $E_{a,z}$ is the activation energy; and $R_0 = 8.314\,J\,K^{-1}\,mol^{-1}$ is the ideal gas constant.

The normalized concentration c_z can be determined by

$$c_z = c_{z,0} - \int_0^t \frac{dc_z}{dt}\, d\tau \tag{10}$$

The regeneration of the SEI is proportional to the lithium-solvent reaction occurring at the anode. This regeneration can counterbalance the SEI decomposition, as described by [26]

$$\frac{dc_{SEI}}{dt} = \frac{dc_{SEI\text{-}d}}{dt} - \frac{dc_{SEI\text{-}r}}{dt} \tag{11}$$

$$\frac{dc_{SEI\text{-}r}}{dt} = K_{SEI\text{-}r} \cdot \frac{dc_{an}}{dt} \tag{12}$$

$\frac{dc_{SEI\text{-}d}}{dt}$ represents the SEI decomposition rate, which can be calculated using Equation (9). The term $\frac{dc_{SEI\text{-}r}}{dt}$ signifies the SEI regeneration rate, with $K_{SEI\text{-}r}$ as a proportional fitting factor, set to $K_{SEI\text{-}r} = 6$ [26].

The anode decomposition rate is extended with a correction factor that includes the regeneration of the SEI [37,38]:

$$\frac{dc_{an}}{dt} = A_{an} \cdot c_{an}^{n_{an}} \cdot \exp\left(-\frac{E_{a,an}}{RT}\right) \cdot \exp\left(-\frac{c_{SEI}}{c_{SEI,0}}\right) \tag{13}$$

The values of the kinetic parameters for the decomposition submodel are listed in Table 3. It was assumed that the other cell properties remain unchanged in the cells investigated with two different cathode stoichiometries.

Table 3. Kinetic parameters for decomposition reactions.

z	Δh_z (J g^{-1})	$c_{z,0}$ (-)	$n_{z,1}$(-)	$n_{z,2}$(-)	A_z (s^{-1})	$E_{a,z}$ (J mol^{-1})	Ref.
SEI	257	0.15	5.5	0	5.32×10^{10}	9.64×10^4	[21]
CEI	257	0.15	5.5	0	1.21×10^{13}	1.03×10^5	[21]
Anode	1714	1	1	0	0.038 (T < 260 °C) 5 (T > 260 °C)	3.3×10^4	[26]
Separator	−190	1	1	0	1.5×10^{50}	4.2×10^5	[26]
Cathode$_{NMC111}$	154	0.04	1	1	2.25×10^{14}	7.9×10^5	[38]
Cathode$_{NMC811}$	1596	0.33	1	1	2.1×10^{17}	2.499×10^5	[39]
Binder	1500	1	1	0	1.077×10^{12}	1.5885×10^5	[35]
Electrolyte	800	1	1	0	3×10^{15}	1.7×10^5	[23]

2.2. Pressure and Venting Model

The pressure model is closely linked to the previously described thermal model, as it relies on temperature and decomposition reaction rates provided by the thermal model to calculate the pressure generated within the cell.

As illustrated in Figure 3, when the system's temperature increases due to generated or external heat, the internal pressure rises. This increase is primarily due to the evaporation of the electrolyte and the gases generated from decomposition reactions [16]. The internal pressure continues to rise with increased temperatures until a critical pressure threshold is reached, at which point the cell vents, releasing the accumulated gases and reducing the internal pressure [5]. Therefore, it is necessary to consider two distinct pressure regimes: the static pressure before cell venting and the dynamic pressure following the venting and gas outflow.

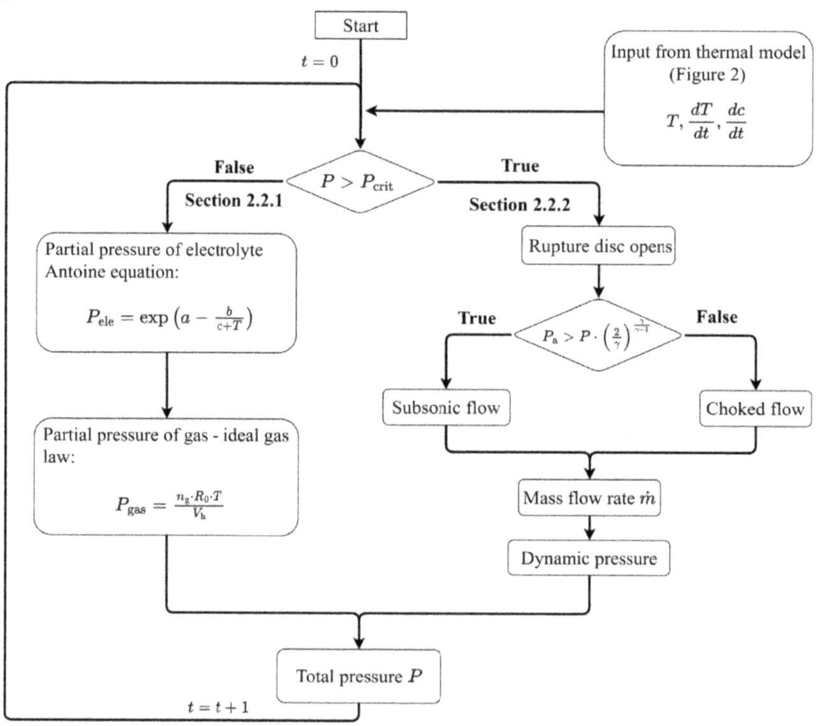

Figure 3. Flowchart for calculating pressure development in lithium-ion cells.

2.2.1. Internal Pressure Development Prior to Cell Opening

The initial increase in pressure is caused by the vaporized electrolyte according to He et al. [21], which can be calculated using the Antoine equation. The Antoine equation is a mathematical formula that relates the saturated vapor pressure of pure substances to temperature [40]:

$$P_{\text{ele}} = \exp\left(a - \frac{b}{c+T}\right) \tag{14}$$

where a, b, and c are fitting parameters used to align with pressure measurement data from experiments conducted before the onset of decomposition reactions.

The pressure increase due to gas production from decomposition reactions can be calculated using the ideal gas equation. He et al. [21] developed a simplified model to predict the pressure increase within the cell, taking into account gas generation primarily from the decomposition of the SEI and CEI layers. These decomposition reactions are the main contributors to internal gas production until the venting event. The partial pressure of the gas is calculated using Equation (15):

$$P_{\text{gas}} = \frac{n_{\text{gas}} R_0 T}{V_{\text{h}}} \tag{15}$$

where n_{gas} is the amount of generated gas in moles, R_0 is the universal gas constant, and V_{h} is the headspace volume of the cell, which accounts for 7% of the cell's total volume [5].

The quantity of gas generated, n_{gas}, is derived from Equation (16):

$$n_{\text{gas}} = \frac{m_{\text{gas,SEI}} + m_{\text{gas,CEI}}}{M_{\text{gas}}} \tag{16}$$

where $m_{\text{gas,SEI}}$ and $m_{\text{gas,CEI}}$ represent the mass of gas produced from SEI and CEI decomposition reactions, respectively, and M_{gas} is the molar mass of the generated gas.

The masses $m_{\text{gas,SEI}}$ and $m_{\text{gas,CEI}}$ can be determined by Equations (17)–(20):

$$c_{\text{gas,SEI}} = c_{\text{SEI,0}} - c_{\text{SEI}} \tag{17}$$

$$c_{\text{gas,CEI}} = c_{\text{CEI,0}} - c_{\text{CEI}} \tag{18}$$

$$m_{\text{gas,SEI}} = m_{\text{gas,SEI,0}} \cdot c_{\text{gas,SEI}} \tag{19}$$

$$m_{\text{gas,CEI}} = m_{\text{gas,CEI,0}} \cdot c_{\text{gas,CEI}} \tag{20}$$

where c_{SEI} and c_{CEI} are calculated using Equation (10); $m_{\text{gas,SEI,0}}$ and $m_{\text{gas,CEI,0}}$ represent the total masses of gases generated from SEI and CEI decomposition, accounting for 4.7% and 0.8% of the reactants' masses, respectively [21].

The overall internal pressure inside the cell is the sum of the pressure arising from the evaporated electrolyte and the pressure due to gas generated from decomposition reactions:

$$P = P_{\text{ele}} + P_{\text{gas}} \tag{21}$$

Table 4 shows the parameters for the pressure development in the cell prior to venting.

Table 4. Fitting and calculated parameters used to calculate the internal pressure in the cell prior to venting.

Parameter	Symbol	Value	Unit
Antoine constant	a	0.97 * for NMC111, 0.34 for NMC811	-
Antoine constant	b	150.49	K
Antoine constant	c	−279.85	K
Headspace volume	V_{h}	21.5	cm^3

* The value 0.97 for NMC111 takes into account preheating by 70 °C and the initial pressure value (see Section 3).

2.2.2. Dynamic Pressure Variation During the Venting Process

Wang et al. [22], Ostanel et al. [24], and Kong et al. [25] investigated the pressure dynamics following the venting after the rupture disc opens. They observed that the vapor–liquid equilibrium is disrupted upon venting, rendering the previously established equations in Section 2.2.1 invalid. The change in internal cell pressure can then be determined by differentiating the two pressure components in Equation (21) with respect to time:

$$\frac{dP}{dt} = \frac{dP_{\text{ele}}}{dt} + \frac{dP_{\text{gas}}}{dt} \tag{22}$$

When the rupture disc opens, the mass of electrolyte vapor, as well as the temperature in the cell, changes according to [17]:

$$\frac{dP_{\text{ele}}}{dt} = \frac{R_0 T}{M_{\text{ele}}} \cdot \frac{dm_{\text{ele}}}{dt} + \frac{m_{\text{ele}} R_0}{M_{\text{ele}} V_{\text{h}}} \cdot \frac{dT}{dt} \tag{23}$$

Here, m_{ele} stands for the electrolyte vapor mass in the cell void space; M_{ele} for the molar mass of the electrolyte vapor; and $\frac{dm_{\text{ele}}}{dt}$ is the change in electrolyte vapor mass per time.

$\frac{dm_{\text{ele}}}{dt}$ is calculated by [25]:

$$\frac{dm_{\text{ele}}}{dt} = -\varphi_{\text{ele}} \cdot \frac{dm_{\text{vent}}}{dt} \tag{24}$$

where φ_{ele} is the mass fraction of the electrolyte vapor in the vented mixture; and $\frac{dm_{\text{vent}}}{dt}$ is the mass flow rate of the vent gas. The calculation of $\frac{dm_{\text{vent}}}{dt}$ follows in Section 2.2.3. The integration of Equation (24) provides the instantaneous electrolyte vapor mass m_{ele} in the cell.

Similar calculations are used to simulate the dynamic change of the partial pressure of the generated gases.

$$\frac{dP_{\text{gas}}}{dt} = \frac{R_0 T}{M_{\text{gas}}} \cdot \frac{dm_{\text{gas}}}{dt} + \frac{m_{\text{gas}} R_0}{M_{\text{gas}} V_{\text{h}}} \cdot \frac{dT}{dt} \tag{25}$$

The change in the gas quantity in the cell $\frac{dm_{\text{gas}}}{dt}$ can be determined by the difference between the gas generated in the cell and the gas flow rate vented through the valve [17,25]:

$$\frac{dm_{\text{gas}}}{dt} = \left(m_{\text{z}} \frac{dc_{\text{z}}}{dt} \right) - \varphi_{\text{gas}} \frac{dm_{\text{vent}}}{dt} \tag{26}$$

where φ_{gas} is the mass fraction of the gas resulting from decomposition reactions in the vent gas. The mass fraction φ of the gases in the mass flow rate should lie within the range of 0 to 1. This fraction represents the proportion of the total mass flow rate attributable to the gaseous components. However, the sum of the mass fractions of the electrolyte vapor and generated gases will inherently be less than 1, as outlined in Table 5. This is because the model only considers the gas phase and does not account for the liquid portion of the electrolyte or solid particles, which are also part of the venting mass.

m_{z} is the mass of gas generated by decomposition reaction and is calculated as follows [17]:

$$m_{\text{z}} = X_i \cdot m_{\text{ventgas}} \tag{27}$$

where X_i is the ratio of the generated heat from the decomposition reaction to the total generated heat; m_{ventgas} is the total mass of the vent gas. The parameters used for calculating the dynamic pressure change are summarized in Table 5, while the ratio of gases produced by the decomposition reactions is detailed in Table 6.

Table 5. Parameters for calculating dynamic pressure changes (Equations (23)–(26)) [17,25].

Parameter	Symbol	Value	Unit
Molar mass of electrolyte vapor	M_{ele}	90.08	$g\,mol^{-1}$
Mass fraction of electrolyte vapor in vent gas	φ_{ele}	0.1	–
Mass fraction of generated gas in vent gas	φ_{gas}	0.2	–

Table 6. Values for the ratio of the generated heat from the decomposition reaction to the total generated heat [26].

z	X_i
SEI	0.03
Anode	0.28
Cathode	0.05
Electrolyte	0.14

2.2.3. Mass Flow Rate of Vent Gas

The mass flow rate $\frac{dm_{vent}}{dt}$ is calculated as follows, as proposed by Mao et al. [18]:

$$\frac{dm_{vent}}{dt} = \frac{uP_{vent}A_{vent}M_{gas}}{RT_{vent}} \tag{28}$$

Here, u represents the gas velocity; A_{vent} is the cross-sectional area of the burst opening, which is provided in the data sheet as $A_{vent} = 2.178$ cm^2; P_{vent} is the pressure at the exit of the rupture disc opening in bar; and T_{vent} is the temperature at the exit of the opening. These parameters can be calculated using [5,24,25,41]

$$P_{vent} = \frac{P}{\left(1 + \frac{\gamma-1}{2}\mathrm{Ma}^2\right)^{\frac{\gamma}{\gamma-1}}} \tag{29}$$

$$T_{vent} = \frac{T}{\left(1 + \frac{\gamma-1}{2}\mathrm{Ma}^2\right)} \tag{30}$$

$$u = \mathrm{Ma}\sqrt{\frac{\gamma R T_{vent}}{M_{gas}}} \tag{31}$$

$$\mathrm{Ma} = \begin{cases} 0 & \text{if } P < P_{crit} \quad \text{No flow} \\ \sqrt{\frac{2}{\gamma-1}\left[\left(\frac{P}{P_a}\right)^{\frac{\gamma-1}{\gamma}} - 1\right]} & \text{if } P_a > P\left[\frac{2}{\gamma+1}\right]^{\frac{\gamma-1}{\gamma}} \quad \text{Subsonic} \\ 1 & \text{if } P_a \leq P\left[\frac{2}{\gamma+1}\right]^{\frac{\gamma-1}{\gamma}} \quad \text{Choked flow} \end{cases} \tag{32}$$

P_a is the ambient pressure; γ is the heat capacity ratio $C_{p,gas}/C_{v,gas}$; and Ma is the Mach number, ranges between 0 and 1.

3. Experiments

The experimental validation of the thermal runaway model was carried out through abuse tests conducted with prismatic NMC111 and NMC811 cells. The experimental setups and conditions varied to evaluate the model's flexibility and applicability across different scenarios. To simulate real-world conditions, these tests were performed in a non-isolated test chamber. In all test scenarios presented in this paper, a constant heat ramp was used to trigger the thermal runaway of the cells.

As shown in Figure 4, the experiment setup of the cell consists of (1) a mounting and heating plate fixed on the sides of the cell, (2) temperature measuring sensors fixed on all sides of the cell (T_1–T_4) and on the rupture disc (labeled as "vent") T_{V1}, (3) a temperature control sensor $T_{Control}$ directly connected to the cell to regulate heat input and maintain a consistent heating rate, and (4) a pressure sensor inserted through the seal pin.

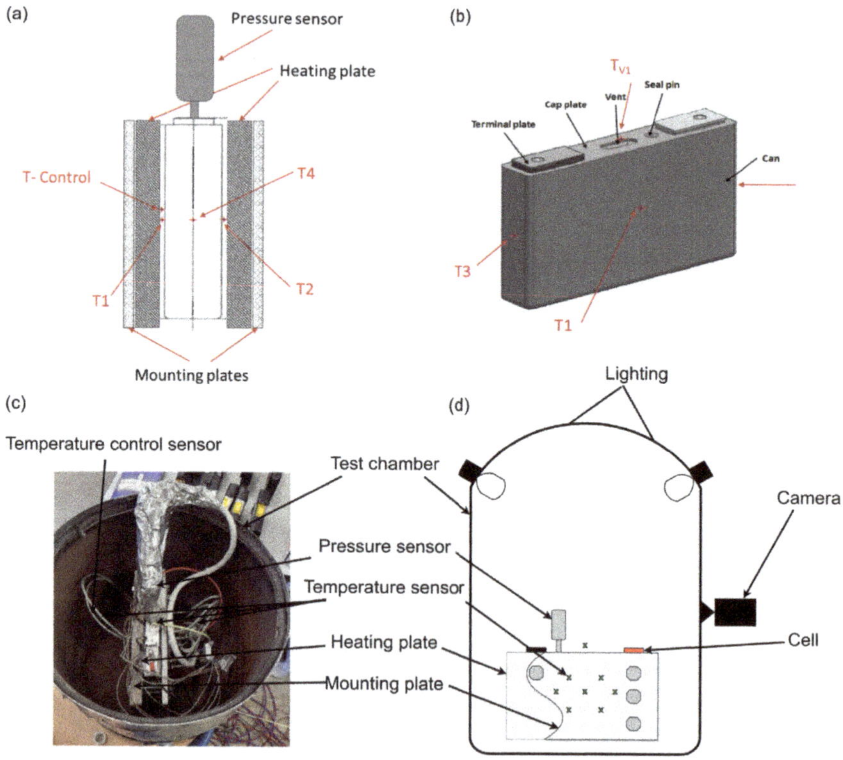

Figure 4. Experimental setup and test chamber configuration: (**a**) Side view of the battery cell illustrating the layout of the pressure sensor, temperature sensors, and heating plates; (**b**) tested prismatic cell with locations of temperature sensors and other key components; (**c**) real experimental chamber setup showcasing the arrangement of different temperature sensors and other elements; (**d**) schematic representation of the test chamber.

Conditionsand Boundaries of the Experiments

The experimental conditions and boundaries are summarized in Table 7. In the first experiment, conducted by Orozav [19], an NMC111 cell with dimensions of 148 mm × 92 mm × 26.5 mm and a capacity of 37 Ah was utilized. While international abuse testing standards for lithium-ion batteries in electric vehicles typically recommend heating rates of 5–10 °C/min [42], a lower heating rate of 0.5 K/min was employed to ensure uniform temperature distribution by heating the cell from both sides. Prior to the commencement of the heating phase, the cell was preheated to 70 °C and maintained at this elevated temperature for a duration of 30 min. This preheating phase ensured a homogeneous temperature distribution and chemical stability, thereby establishing a controlled initial condition. Given the new condition of the cell and its specification allowing storage up to 70 °C, it is presumed that minimal or negligible decomposition reactions had occurred. As can be seen in Section 4.1.2, preheating caused a slight increase in pressure. The heating process was subsequently initiated with continuous monitoring of the cell temperature and internal pressure. This experimental setup aimed to validate the model's response to different initial conditions and heating profiles, demonstrating the model's flexibility under varying experimental scenarios.

In the second experiment, an NMC811 cell with the with dimensions of 148 mm × 94 mm × 26.5 mm and a capacity of 50 Ah was employed. The cell was preheated to 25 °C and maintained at this temperature for 30 min to ensure a homogeneous temperature distribution and chemical stability prior to the test. Following this stabilization period, the cell was subjected to asymmetric heating from one side at a rate of 10 K/min. This experimental condition aimed to assess the model's performance and simulation results under a different heating profile and configuration compared to the first experiment. Furthermore, this condition more closely simulates real-world scenarios where heat might propagate from an adjacent cell in a battery pack. Additionally, the high SOC of 95% represents a worst-case scenario for a commercially used cell. Continuous monitoring of the cell's temperature and internal pressure was conducted throughout the heating process, allowing for a comprehensive evaluation of the model under these specific experimental conditions.

Table 7. Comparison of the experimental conditions for the NMC111 and NMC811 experiments.

Parameter	NMC111	NMC811
Capacity	37 Ah	50 Ah
State of Charge (SoC)	80%	95%
Heating rate	0.5 K min^{-1}	10 K min^{-1}
Heating sides	Both sides	One side
Initial temperature	70 °C	25 °C

4. Results and Discussion

In this section, a comprehensive analysis is presented of the thermal and pressure dynamics occurring during thermal runaway events in the examined NMC111 and NMC811 lithium-ion cells. Controlled experiments and simulations were employed to evaluate the accuracy and reliability of the simulation models. Focus is placed on comparing temperature and pressure profiles across different experimental setups to determine the models' effectiveness in replicating the cell behavior during thermal instability events.

4.1. Discussion of NMC111 Experimental and Simulation Results

4.1.1. Temperature Profile of the NMC111

Figure 5 displays the temperature profile of the tested NMC111 cell, alongside the simulation results. As previously detailed, the cell was subjected to heating from an initial temperature of 70 °C at a rate of 0.5 K/min. The temperature increases gradually due to the low heating rate until the TR event occurs. During the experiment, a temperature drop was observed at the venting event. This drop is attributed to the dissipation of heat with the venting gas, where a part of the heat was carried away by the expelled gas, leading to a noticeable decrease in temperature. However, this temperature drop was not replicated in the simulation due to the model's assumption that the heat flow caused by venting is compensated by a higher external heating rate (ideal heating ramp, dT/dt is constant)).

When comparing the experimental data with the simulation, it is evident that TR occurs eleven seconds earlier in the model than in the experiment. A notable difference is also observed in the maximum temperature during TR. The simulation shows a higher maximum temperature compared to the experimental data, with a discrepancy of 80 °C. Several factors could contribute to this difference: the simulated temperature is an average of readings from four temperature sensors (T_1, T_2, T_3, and T_4), as shown in Figure 4, excluding the temperature at the rupture disc. Furthermore, the model does not consider the heat dissipation due to solid and fluid particles expelled during the second venting event of the TR. Exact comparison values are listed in Table 8.

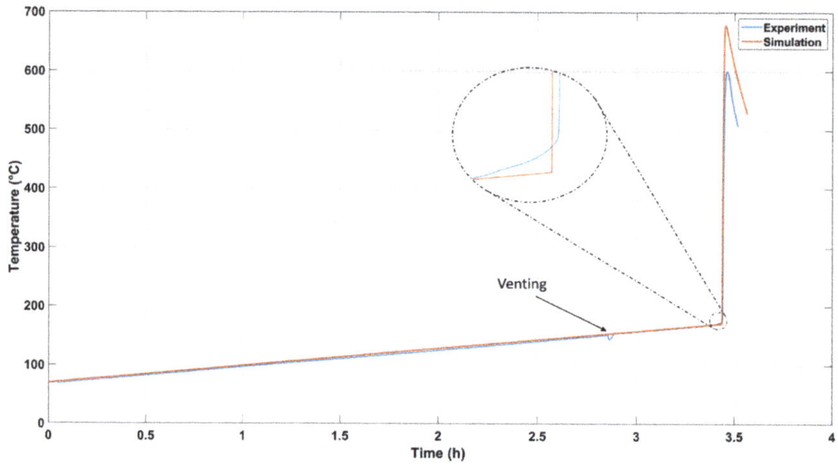

Figure 5. Comparison between measured and simulated temperature of the first experiment with NMC111 and a heating rate of 0.5 K/min (see parameters in Table 7).

Table 8. Comparison of the experimental data and simulation results of the experiments with NMC111.

Parameter	Experiment	Simulation
Time of TR t_{TR}	3 h and 26 min	3 h and 26 min
Maximal Temperature by TR T_{max}	600 °C	680 °C
Time of first venting t_{vent}	2 h and 51 min	2 h and 50 min
Temperature of first venting T_{vent}	152 °C	153.5 °C
Pressure increase by TR P_{TR}	N/A	0.74 bar

When evaluating the experimentally determined maximum temperature, great attention must be paid to the test conditions, as the position and dynamics of the temperature sensors can result in large differences, as Section 4.2.1 also shows. For this reason, results from different sources are often difficult to compare if the position and type of sensors and the test conditions are not exactly the same.

4.1.2. Pressure Profile of the NMC111

Figure 6 shows the pressure profile of the NMC111 cell during the experiment and its comparison with the simulation results. The initial pressure of 0.26 bar is caused by preheating the cell to 70 °C and maintaining this temperature for 30 min. Following this preheating phase, the temperature rise causes the pressure inside the cell to increase exponentially due to the evaporation of the electrolyte and the generation of gas until it reaches a critical pressure of 4.46 bar, prompting the cell to vent. In the experimental data, this critical pressure was reached after 2 h and 51 min at a temperature of 152 °C, whereas in the simulation, the predicted first venting occurred after 2 h and 50 min at a temperature of 153.5 °C, demonstrating a minimal discrepancy in both time and temperature.

Following the venting or opening of the cell, the pressure rapidly decreases to ambient levels due to the release of accumulated gases, which causes a corresponding drop in the temperature curve, as depicted in Figure 5. The pressure remains at ambient levels until the occurrence of TR. During the experimental procedure, the high temperatures during the TR process rendered the pressure sensor ineffective, preventing pressure measurements during this period. In contrast, the model predicts a significant pressure increase during TR, attributed to the rapid and substantial gas generation. The model indicates that the pressure increases again during TR, with a predicted rise of approximately 0.75 bar due to the substantial amount of gas generated. This is similar to the results shown by Zhang et al. [17], Wang et al. [22], Ostanek et al. [24], and Kong et al. [25].

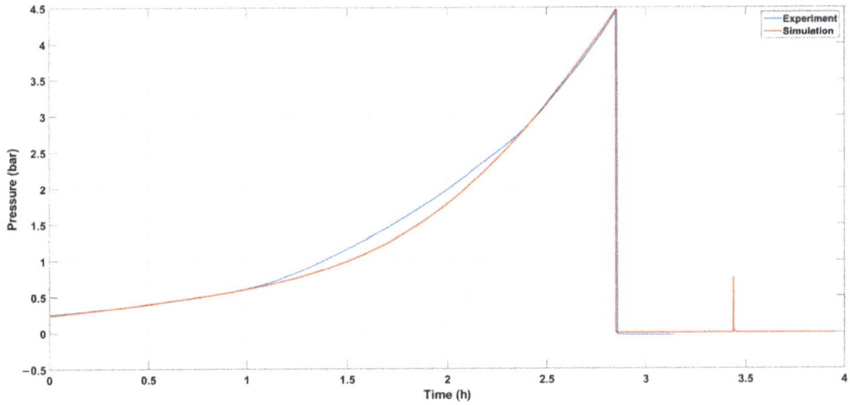

Figure 6. Comparison between measured and simulated pressure profiles of the NMC111 cell, highlighting pressure dynamics during thermal runaway, which were not measured experimentally.

The disparity in the pressure profiles between the experimental and simulated curves from 1 to 2.5 h can be attributed to several factors. Firstly, the 30-min preheating phase was not accounted for in the model, affecting the internal dynamics of the cell and leading to differences in the pressure build-up within this time interval. Additionally, small variations in experimental conditions, such as the rate of electrolyte evaporation and gas generation, may have contributed to the observed differences in pressure development between the two profiles. These discrepancies are considered acceptable given that many parameters in the model are empirically determined (see, for instance, Equation (14) concerning the Antoine parameters a, b, c and the mass fractions φ_{ele} and φ_{gas}). Moreover, the calculated root mean square error (RMSE) of 0.18 indicates a reasonably close fit between the simulated and experimental data. This RMSE value demonstrates the robustness of the model despite the simplifications and supports its reliability in predicting the pressure dynamics.

The results of the model compared to the experiment with NMC111 are listed in Table 8.

4.2. Discussion of NMC811 Experimental and Simulation Results

4.2.1. Temperature Profile of the NMC811

Figure 7 compares the simulated and measured temperature profiles of the NMC811 cell. The cell was heated from an initial temperature of 25 °C with a heating rate of 10 K/min from one side until TR occurred. Similar to the first experiment, the simulated temperature represents the average temperature recorded by various sensors positioned on the sides of the cell, as illustrated in Figure 4. The experimental data indicates an oscillating temperature curve due to the high-power heating plate used in this experiment, which was necessary to maintain the intended heating rate. As a result, no temperature drop due to the first venting was detected. Venting was observed shortly before the TR, occurring after approximately 22 min, as noted in Figure 8. This was confirmed through both visual observation during the experiment and pressure measurements.

A heating rate of 7 K/min is used in the model for the simulations, as the average temperature was simulated in the lumped thermal model. This means the heating rate reflects the temperature in the middle of the cell rather than at the sides, leading to differing internal temperature distributions. The non-linearity in the temperature profile is due to heating from one side and the thermal conductivity properties inside the cell.

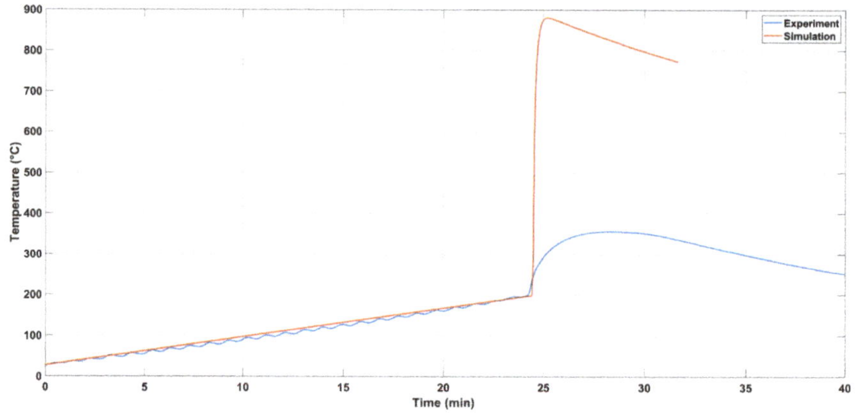

Figure 7. Comparison between measured and simulated temperature of abuse experiment with NMC811 and heating rate of 10 K/min.

The measured data show that TR was triggered 13 s earlier in the experiment compared to the simulation. The maximum temperature reached during TR in the simulation was 880 °C, whereas it was 350 °C in the measured data. This discrepancy can be attributed to the temperature sensors used, which likely have a larger diameter and a slower response time. Nevertheless, despite the high heating rate and the inhomogeneous temperature distribution in the cell, the time of the TR is well matched. The quality of the simulation is even better described by the pressure development, see Section 4.2.2.

4.2.2. Pressure Profile of the NMC811

As depicted in Figure 8, the internal pressure of the NMC811 cell increases exponentially with increasing temperature until it reaches 4.7 bar, at which point the rupture disc opens. Similar to the behavior observed in the NMC111 experiments, the pressure drops abruptly to 0 bar upon venting. In this particular experiment, a high-temperature-resistant sensor was utilized, recording a pressure of 1.6 bar during TR. The simulation, on the other hand, calculated a pressure of 1.25 bar.

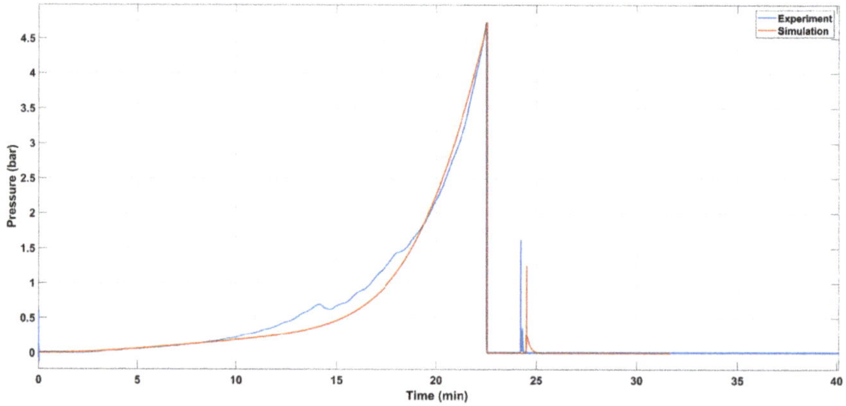

Figure 8. Comparison of measured and simulated pressure profiles for the NMC811 cell, highlighting the timing of venting events and pressure increase during thermal runaway.

The calculated RMSE for the pressure profiles is 0.23, suggesting a moderate level of discrepancy between the simulation and the experimental data. This RMSE value reflects the simulation's fairly good performance in capturing the general trend and dynamics of pressure build-up within the cell. Notably, the simulation closely aligns with experimental observations, with the cell venting occurring only three seconds earlier than predicted. Additionally, the venting temperature was consistently captured at 185 °C in both the simulation and experimental scenarios.

The results of the model, as compared to the experiment with NMC811, are summarized in Table 9.

Table 9. Comparison of the experimental and simulation results of the experiments with NMC811.

Parameter	Experiment	Simulation
Time of TR (t_{TR})	24 min and 14 s	24 min and 27 s
Maximal Temperature during TR (T_{max})	350 °C	880 °C
Time of First Venting (t_{vent})	22 min and 30 s	22 min and 33 s
Temperature of First Venting (T_{vent})	185 °C	185 °C
Pressure during TR (P_{TR})	1.6 bar	1.25 bar

4.3. Additional Model Outputs

The model offers a range of additional outputs, providing insights into various thermal processes within the cell during a TR event. These outputs include the total generated heat, heat produced from the decomposition of different cell components, and dissipated heat, among others. Additionally, the model can simulate the cell voltage during TR events, offering a comprehensive understanding of the cell's thermal behavior and electrical response under extreme conditions.

One key focus, as shown in Figure 9, is the self-heating due to decomposition reactions, which begins with the breakdown of the SEI. With the increase in temperature, additional decomposition reactions are triggered, further contributing to self-heating and driving it toward a critical point where the separator begins to melt. The melting of the separator is an endothermic process, consuming energy and thus leading to a temporary decrease in apparent self-heating. However, these negative values during separator melting are not visible on the self-heating curve due to the limitations of the logarithmic scale used, a situation illustrated in Figure 9. Despite this, the heat input curve does reflect this phenomenon. It shows an increase in the supplied heat power to compensate for the energy absorbed by the separator's melting, ensuring the continued progression of thermal processes within the cell, maintaining the required heating rate of 7 K/min, as detailed in Section 4.2.1.

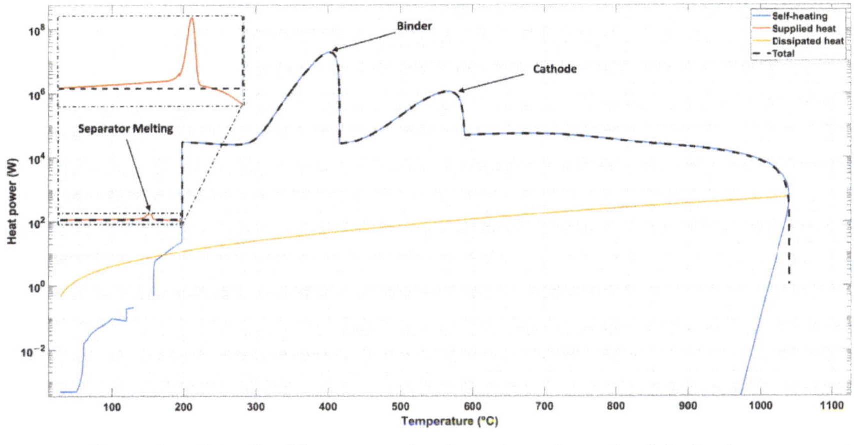

Figure 9. Illustration of simulated heat generation dynamics and associated dissipation processes in the NMC811 cell model.

The self-heating continues to escalate until the onset of thermal runaway (TR), where the generation power surges dramatically due to heat produced from redox reactions and potential short circuits, as described in Section 2.1. This rapid increase in thermal power is marked by two distinct peaks within the temperature range of 300–600 °C, which correspond to the decomposition of the binder and cathode materials, respectively. The heating power associated with these specific decomposition events is illustrated in Figure 10, offering a detailed view of the contributions from each component during these key temperature phases. The dissipated heat, calculated using Equations (7) and (8), shows a gradual increase with temperature elevation.

Figure 9 also clearly illustrates the model's constant total heat generation up until the onset of TR, correlating with the intended constant heating rate. The supplied heat, highlighted in the zoomed excerpt, is dynamically adjusted over time to compensate for dissipated heat and the energy consumed during separator melting, ensuring the heating rate remains consistent. After the separator melting, self-heating intensifies, reducing the amount of external power needed, which is reflected in the decline of the supplied power curve until it eventually ceases at TR. This detailed breakdown highlights the pivotal moment when self-heating exceeds dissipated heat, signifying that the cell will autonomously continue to increase in temperature, progressing toward TR, even in the absence of external heat input. This underscores the need for active cooling measures to prevent uncontrolled temperature rises and provides an estimation of the cooling power required to avert TR.

Figure 10 illustrates the simulated heat generation power and cumulative energy derived from decomposition reactions of various cell components at different temperatures. The figure identifies the key components responsible for heat generation and their corresponding temperature ranges. Figure 10a displays the heat power generated by each reaction, while Figure 10b shows the cumulative heat in joules. Self-heating initiates with the decomposition of the SEI, followed by anode decomposition starting around 120 °C. The decomposition of the electrolyte begins at approximately 140 °C, contributing significantly to heat generation, particularly dominant between 600 °C and 700 °C. At temperatures exceeding 200 °C, decomposition reactions from the cathode and binder further amplify the self-heating process, underscoring their roles in elevating cell temperatures during TR event.

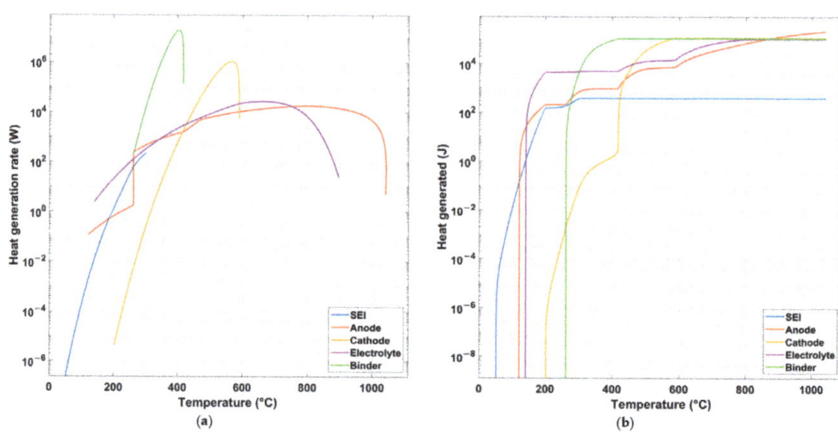

Figure 10. Simulated (**a**) heat generation power and (**b**) cumulative energy resulting from decomposition reactions of cell components, highlighting their roles across temperature ranges.

Additionally, the model can calculate the cell's voltage response to temperature changes during an abuse test, as depicted in Figure 11 for the example with the NMC811 cell. The cell starts with a state of charge (SoC) of 95% and an initial voltage of 4.12 V.

During the experiment, the voltage experiences a slight decrease of 0.01 V before the cell undergoes a short circuit. In contrast, the simulation indicates a voltage drop of 0.13 V before the short circuit. Following the onset of thermal runaway and the short circuit, the voltage sharply declines to 0 V but momentarily rebounds before ultimately falling to 0 V. Although the model does not exactly match the experimental data, it accurately captures the overall trend, including the fluctuation in the voltage observed experimentally. This behavior can be attributed to a decrease in internal cell resistance due to the temperature rise [43].

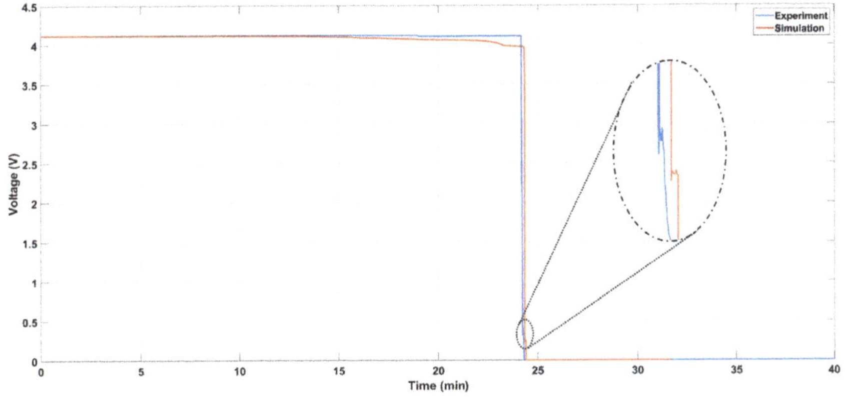

Figure 11. Simulated voltage response of an NMC811 cell during thermal abuse testing.

5. Conclusions

In this study, a comprehensive simulation model was developed to replicate the temperature and pressure development within prismatic lithium-ion cells, specifically focusing on NMC111 and NMC811 chemistries. The flexibility and accuracy of the model were rigorously evaluated under varied experimental conditions, including variations in heating rates, initial temperatures, and states of charge.

The thermal model incorporated into the simulation is based on the sequential decomposition reactions of key cell components, effectively capturing the heat production dynamics of a cell undergoing thermal abuse. This approach allowed for detailed replication of the cell's behavior during thermal runaway, considering both the heat generated internally and the heat dissipated to the environment.

Aligned with the thermal model, a pressure sub-model was developed to simulate the pressure rise inside the cell. This was attributed to both the evaporation of the electrolyte and the generation of gases from decomposition reactions. The simulation effectively modeled the critical pressure points leading to cell venting, accurately reflecting the experimental observations.

Upon validation, the model exhibited a slight discrepancy in the timing of thermal runaway and the maximum temperatures achieved compared to experimental data. Such differences have been attributed to the inherent variability in sensor responses and experimental conditions. Nonetheless, the overall trends observed in both temperature and pressure profiles were closely mirrored by the simulation results, demonstrating the robustness and reliability of the model.

Future enhancements could involve integrating sub-models like a cooling model to simulate thermal management strategies and extending the model to assess propagation phenomena within battery packs.

In conclusion, this simulation model offers a valuable tool for predicting the thermal and pressure behaviors of lithium-ion cells under various stress scenarios, enhancing the understanding of cell safety and reliability. Future work could refine the model by including additional variables such as cell aging and the exploration of thermal runaway propagation among cells to further enhance its predictive capabilities.

Author Contributions: Conceptualization, M.A., T.R. and R.B.; methodology, M.A. and K.K.; software, M.A. and K.K.; validation, M.A. and T.R.; formal analysis, M.A.; investigation, M.A. and T.R.; resources, R.B.; data curation, M.A.; writing—original draft preparation, M.A.; writing—review and editing, M.A., R.B., J.K. and I.H.; visualization, M.A.; supervision, R.B. and I.H.; project administration, R.B.; funding acquisition, R.B. All authors have read and agreed to the published version of the manuscript.

Funding: Parts of this research were funded by the Federal Ministry for Economic Affairs and Energy of Germany in the project RiskBatt (project number: 03EI3010A). Other parts were funded by the Federal Ministry of Education and Research of Germany in the project VentBatt (project number: 03XP0535A).

Data Availability Statement: The data from the experiments are publicly accessible on the Zenodo platform, https://doi.org/10.5281/zenodo.13981390.

Acknowledgments: The authors acknowledge the support of the Open Access Publishing Fund of Clausthal University of Technology.

Conflicts of Interest: The authors declare no conflicts of interest.

Nomenclature

Variables	
$\frac{dc}{dt}$	decomposition rate (s^{-1})
$\dot{Q}$	flow of heat rate (W)
$\frac{dm_{ele}}{dt}$	rate of mass change ($kg\,s^{-1}$)
$\frac{dm_{vent}}{dt}$	mass flow rate of the vent gas ($kg\,s^{-1}$)
$\frac{dP}{dt}$	pressure change rate ($Pa\,s^{-1}$)
$\frac{dT}{dt}$	rate of temperature change ($K\,s^{-1}$)
A_z	reaction frequency factor (s^{-1})
A_{cell}	cell dissipation area with ambient (m^2)
c	normalized concentration (−)
c_p	specific heat capacity ($J\,kg^{-1}\,K^{-1}$)
E_a	activation energy ($J\,mol^{-1}$)
h_{conv}	convection heat transfer coefficient ($W\,m^{-2}\,K^{-1}$)
M	molar mass ($kg\,mol^{-1}$)
m	mass (kg)
Ma	mach number (−)
n_z	reaction order (−)
P	pressure (Pa)
R	cell resistance (Ω)
R_0	gas constant ($J\,K^{-1}\,mol$)
T	temperature (K)
u	velocity ($m\,s^{-1}$)
V	voltage of the cell (V)
V_h	headspace volume (m^3)
X_i	ratio of the generated heat from z reaction to the total generated heat (−)
Greek	
ΔH	enthalpy (J)
Δh	specific enthalpy ($J\,kg^{-1}$)
Δt	average short circuit time (s)
γ	heat capacity ratio (−)
σ	surface emissivity ($W\,m^{-2}\,K^{-4}$)
ε	surface emissivity (−)
φ	mass fraction in the vent gas (−)
ς	efficiency factor used to correct the non-ohmic heat generation (−)

Subscripts

0	initial value
cat&an	oxidation and reduction reactions between cathode and anode
diss	heat dissipation
ISC	internal short circuit
a	ambient conditions
an	anode
ele	electrolyte
gas	parameters for generated gas from decomposition
SEI-r	SEI-regeneration
vent	vent gas
z	indicates the variable corresponds to reactant
Acronyms	
ARC	accelerating rate calorimeter
CEI	cathode electrolyte interphase
LFP	lithium iron phosphate
NMC	lithium nickel manganese cobalt oxides
PE	polyethylene
PP	polypropylene
SEI	solid electrolyte interphase
SoC	state of charge

References

1. Şen, M.; Özcan, M.; Eker, Y. A Review on the Lithium-Ion Battery Problems Used in Electric Vehicles. *Next Sust.* **2024**, *3*, 100036. [CrossRef]
2. Ramadesigan, V.; Northrop, P.W.C.; De, S.; Santhanagopalan, S.; Braatz, R.D.; Subramanian, V.R. Modeling and Simulation of Lithium-Ion Batteries from a Systems Engineering Perspective. *J. Electrochem. Soc.* **2012**, *159*, R31. [CrossRef]
3. Christensen, P.A.; Anderson, P.A.; Harper, G.D.; Lambert, S.M.; Mrozik, W.; Rajaeifar, M.A.; Wise, M.S.; Heidrich, O. Risk Management over the Life Cycle of Lithium-Ion Batteries in Electric Vehicles. *Renew. Sustain. Energy Rev.* **2021**, *148*, 111240. [CrossRef]
4. Xu, C.; Fan, Z.; Zhang, M.; Wang, P.; Wang, H.; Jin, C.; Peng, Y.; Jiang, F.; Feng, X.; Ouyang, M. A Comparative Study of the Venting Gas of Lithium-Ion Batteries during Thermal Runaway Triggered by Various Methods. *Cell Rep. Phys. Sci.* **2023**, *4*, 101705 . [CrossRef]
5. Coman, P.T.; Mátéfi-Tempfli, S.; Veje, C.T.; White, R.E. Modeling Vaporization, Gas Generation and Venting in Li-Ion Battery Cells with a Dimethyl Carbonate Electrolyte. *J. Electrochem. Soc.* **2017**, *164*, A1858. [CrossRef]
6. Klink, J.; Hebenbrock, A.; Grabow, J.; Orazov, N.; Nylén, U.; Benger, R.; Beck, H.P. Comparison of Model-Based and Sensor-Based Detection of Thermal Runaway in Li-Ion Battery Modules for Automotive Application. *Batteries* **2022**, *8*, 34. [CrossRef]
7. E, J.; Xiao, H.; Tian, S.; Huang, Y. A Comprehensive Review on Thermal Runaway Model of a Lithium-Ion Battery: Mechanism, Thermal, Mechanical, Propagation, Gas Venting and Combustion. *Renew. Energy* **2024**, *229*, 120762. [CrossRef]
8. Hu, X.; Gao, F.; Xiao, Y.; Wang, D.; Gao, Z.; Huang, Z.; Ren, S.; Jiang, N.; Wu, S. Advancements in the Safety of Lithium-Ion Battery: The Trigger, Consequence and Mitigation Method of Thermal Runaway. *Chem. Eng. J.* **2024**, *481*, 148450. [CrossRef]
9. Qiu, M.; Liu, J.; Cong, B.; Cui, Y. Research Progress in Thermal Runaway Vent Gas Characteristics of Li-Ion Battery. *Batteries* **2023**, *9*, 411. [CrossRef]
10. Wang, Q.; Mao, B.; Stoliarov, S.I.; Sun, J. A Review of Lithium Ion Battery Failure Mechanisms and Fire Prevention Strategies. *Prog. Energy Combust. Sci.* **2019**, *73*, 95–131. [CrossRef]
11. Feng, X.; Ouyang, M.; Liu, X.; Lu, L.; Xia, Y.; He, X. Thermal Runaway Mechanism of Lithium Ion Battery for Electric Vehicles: A Review. *Energy Storage Mater.* **2018**, *10*, 246–267. [CrossRef]
12. Feng, X.; Sun, J.; Ouyang, M.; He, X.; Lu, L.; Han, X.; Fang, M.; Peng, H. Characterization of Large Format Lithium Ion Battery Exposed to Extremely High Temperature. *J. Power Sources* **2014**, *272*, 457–467. [CrossRef]
13. Li, Y.; Pu, H.; Wei, Y. Polypropylene/Polyethylene Multilayer Separators with Enhanced Thermal Stability for Lithium-Ion Battery via Multilayer Coextrusion. *Electrochim. Acta* **2018**, *264*, 140–149. [CrossRef]
14. Zhang, Q.K.; Zhang, X.Q.; Yuan, H.; Huang, J.Q. Thermally Stable and Nonflammable Electrolytes for Lithium Metal Batteries: Progress and Perspectives. *Small Sci.* **2021**, *1*, 2100058. [CrossRef]
15. Feng, X.; Zheng, S.; He, X.; Wang, L.; Wang, Y.; Ren, D.; Ouyang, M. Time Sequence Map for Interpreting the Thermal Runaway Mechanism of Lithium-Ion Batteries With $LiNi_xCo_yMn_zO_2$ Cathode. *Front. Energy Res.* **2018**, *6*, 126. [CrossRef]
16. Cai, T.; Tran, V.; Stefanopoulou, A.G.; Siegel, J.B. Modeling Li-ion Battery First Venting Events Before Thermal Runaway. *IFAC-PapersOnLine* **2021**, *54*, 528–533. [CrossRef]
17. Zhang, P.; Lu, J.; Yang, K.; Chen, H.; Huang, Y. A 3D Simulation Model of Thermal Runaway in Li-ion Batteries Coupled Particles Ejection and Jet Flow. *J. Power Sources* **2023**, *580*, 233357. [CrossRef]

18. Mao, B.; Zhao, C.; Chen, H.; Wang, Q.; Sun, J. Experimental and Modeling Analysis of Jet Flow and Fire Dynamics of 18650-Type Lithium-Ion Battery. *Appl. Energy* **2021**, *281*, 116054. [CrossRef]
19. Orazov, N. Methode zur Zustandsbewertung Thermisch Beschädigter Lithium-Ionen-Zellen. Ph.D. Thesis, Clausthal University of Technology, Clausthal-Zellerfeld, Germany, 2024. [CrossRef]
20. Hebenbrock, A.; Orazov, N.; Benger, R.; Schade, W.; Hauer, I.; Turek, T. Innovative Early Detection of High-Temperature Abuse of Prismatic Cells and Post-Abuse Degradation Analysis Using Pressure and External Fiber Bragg Grating Sensors. *Batteries* **2024**, *10*, 92. [CrossRef]
21. He, C.X.; Yue, Q.L.; Chen, Q.; Zhao, T.S. Modeling Thermal Runaway of Lithium-Ion Batteries with a Venting Process. *Appl. Energy* **2022**, *327*, 120110. Simulation und venting. [CrossRef]
22. Wang, G.; Kong, D.; Ping, P.; Wen, J.; He, X.; Zhao, H.; He, X.; Peng, R.; Zhang, Y.; Dai, X. Revealing Particle Venting of Lithium-Ion Batteries during Thermal Runaway: A Multi-Scale Model toward Multiphase Process. *eTransportation* **2023**, *16*, 100237. [CrossRef]
23. Feng, X.; He, X.; Ouyang, M.; Lu, L.; Wu, P.; Kulp, C.; Prasser, S. Thermal Runaway Propagation Model for Designing a Safer Battery Pack with 25 Ah $LiNi_XCo_yMn_zO_2$ Large Format Lithium Ion Batter. *Appl. Energy* **2015**, *154*, 74–91. [CrossRef]
24. Ostanek, J.K.; Li, W.; Mukherjee, P.P.; Crompton, K.R.; Hacker, C. Simulating Onset and Evolution of Thermal Runaway in Li-ion Cells Using a Coupled Thermal and Venting Model. *Appl. Energy* **2020**, *268*, 114972. [CrossRef]
25. Kong, D.; Wang, G.; Ping, P.; Wen, J. A Coupled Conjugate Heat Transfer and CFD Model for the Thermal Runaway Evolution and Jet Fire of 18650 Lithium-Ion Battery under Thermal Abuse. *eTransportation* **2022**, *12*, 100157. [CrossRef]
26. Feng, X.; He, X.; Ouyang, M.; Wang, L.; Lu, L.; Ren, D.; Santhanagopalan, S. A Coupled Electrochemical-Thermal Failure Model for Predicting the Thermal Runaway Behavior of Lithium-Ion Batteries. *J. Electrochem. Soc.* **2018**, *165*, A3748. [CrossRef]
27. Hong, J.; Wang, Z.; Qu, C.; Zhou, Y.; Shan, T.; Zhang, J.; Hou, Y. Investigation on Overcharge-Caused Thermal Runaway of Lithium-Ion Batteries in Real-World Electric Vehicles. *Appl. Energy* **2022**, *321*, 119229. [CrossRef]
28. Ren, D.; Feng, X.; Lu, L.; Ouyang, M.; Zheng, S.; Li, J.; He, X. An Electrochemical-Thermal Coupled Overcharge-to-Thermal-Runaway Model for Lithium Ion Battery. *J. Power Sources* **2017**, *364*, 328–340. [CrossRef]
29. Ren, D.; Liu, X.; Feng, X.; Lu, L.; Ouyang, M.; Li, J.; He, X. Model-Based Thermal Runaway Prediction of Lithium-Ion Batteries from Kinetics Analysis of Cell Components. *Appl. Energy* **2018**, *228*, 633–644. [CrossRef]
30. Chen, J.; Ren, D.; Hsu, H.; Wang, L.; He, X.; Zhang, C.; Feng, X.; Ouyang, M. Investigating the Thermal Runaway Features of Lithium-Ion Batteries Using a Thermal Resistance Network Model. *Appl. Energy* **2021**, *295*, 117038. [CrossRef]
31. Liu, X.; Ren, D.; Hsu, H.; Feng, X.; Xu, G.L.; Zhuang, M.; Gao, H.; Lu, L.; Han, X.; Chu, Z.; et al. Thermal Runaway of Lithium-Ion Batteries without Internal Short Circuit. *Joule* **2018**, *2*, 2047–2064. [CrossRef]
32. Coman, P.T.; Darcy, E.C.; Veje, C.T.; White, R.E. Modelling Li-Ion Cell Thermal Runaway Triggered by an Internal Short Circuit Device Using an Efficiency Factor and Arrhenius Formulations. *J. Electrochem. Soc.* **2017**, *164*, A587. [CrossRef]
33. Kim, G.H.; Pesaran, A.; Spotnitz, R. A Three-Dimensional Thermal Abuse Model for Lithium-Ion Cells. *J. Power Sources* **2007**, *170*, 476–489. [CrossRef]
34. Jindal, P.; Kumar, B.S.; Bhattacharya, J. Coupled Electrochemical-Abuse-Heat-Transfer Model to Predict Thermal Runaway Propagation and Mitigation Strategy for an EV Battery Module. *J. Energy Storage* **2021**, *39*, 102619. [CrossRef]
35. Sun, X.; Xu, X.; Zhao, L.; Jin, H.; Cao, L.; Wang, H. Blocking Analysis of Thermal Runaway of a Lithium-Ion Battery under Local High Temperature Based on the Material Stability and Heat Dissipation Coefficient. *Ionics* **2021**, *27*, 107–122. [CrossRef]
36. Mishra, D.; Jain, A. Multi-Mode Heat Transfer Simulations of the Onset and Propagation of Thermal Runaway in a Pack of Cylindrical Li-Ion Cells. *J. Electrochem. Soc.* **2021**, *168*, 020504. [CrossRef]
37. Peng, P.; Jiang, F. Thermal Safety of Lithium-Ion Batteries with Various Cathode Materials: A Numerical Study. *Int. J. Heat Mass Transf.* **2016**, *103*, 1008–1016. [CrossRef]
38. Kong, D.; Wang, G.; Ping, P.; Wen, J. Numerical Investigation of Thermal Runaway Behavior of Lithium-Ion Batteries with Different Battery Materials and Heating Conditions. *Appl. Therm. Eng.* **2021**, *189*, 116661. [CrossRef]
39. Nguyen, T.T.D. Understanding and Modelling the Thermal Runaway of Li-Ion Batteries. Ph.D. Thesis, Université de Picardie Jules Verne, Amiens, France, 2021.
40. Behr, A.; Agar, D.W.; Jörissen, J.; Vorholt, A.J. (Eds.) Physikalisch-chemische Grundlagen I: Gleichgewichte und Thermodynamik. In *Einführung in die Technische Chemie*; Springer: Berlin/Heidelberg, Germany, 2016; pp. 29–42. [CrossRef]
41. Birch, A.D.; Hughes, D.J.; Swaffield, F. Velocity Decay of High Pressure Jets. *Combust. Sci. Technol.* **1987**, *52*, 161–171. [CrossRef]
42. Ruiz, V.; Pfrang, A.; Kriston, A.; Omar, N.; Van den Bossche, P.; Boon-Brett, L. A Review of International Abuse Testing Standards and Regulations for Lithium Ion Batteries in Electric and Hybrid Electric Vehicles. *Renew. Sustain. Energy Rev.* **2018**, *81*, 1427–1452. [CrossRef]
43. Long, M.K.; Liu, S.; Zhang, G. A Novel Method for Simultaneous Triggering and in Situ Sensing of Internal Short Circuit in Lithium-Ion Cells. *Energy Adv.* **2023**, *2*, 2018–2028. [CrossRef]

batteries

MDPI

Article

A Multi-Encoder BHTP Autoencoder for Robust Lithium Battery SOH Prediction Under Small-Sample Scenarios

Chang Liu [1,*], Shunli Wang [2,*], Zhiqiang Ma [1], Siyuan Guo [1] and Yixiong Qin [1]

1 School of Data Science and Application, Inner Mongolia University of Technology, Hohhot 010080, China; mzq_bim@imut.edu.cn (Z.M.); 20221800715@imut.edu.cn (S.G.); 20231800776@imut.edu.cn (Y.Q.)
2 School of Electric Power, Inner Mongolia University of Technology, Hohhot 010080, China
* Correspondence: 202213505019@imut.edu.cn (C.L.); wangshunli@imut.edu.cn (S.W.)

Abstract

Accurate prediction of the state of health (SOH) in lithium batteries (LiBs) is essential for ensuring operational safety, extending battery lifespan, and enabling effective second-life applications. However, achieving precise SOH prediction under small-sample conditions remains a significant challenge due to inherent variability among battery cells and capacity recovery phenomena, which result in irregular degradation patterns and hinder effective feature extraction. To overcome these challenges, this study introduces an advanced autoencoder-based method specifically designed for SOH prediction in small-sample scenarios. This method employs a multi-encoder structure—comprising token, positional, and temporal encoders—to comprehensively capture the multi-dimensional characteristics of SOH sequences. Furthermore, a BHTP module is integrated to facilitate feature fusion and enhance the model's stability and interpretability. By utilizing a pre-training and fine-tuning strategy, the proposed method effectively reduces computational complexity and the number of model parameters while maintaining high prediction accuracy. The validation of the NASA 18650 lithium cobalt oxide battery dataset under various discharge strategies shows that the proposed method achieves fast convergence and outperforms traditional prediction methods. Compared with other models, our method reduces the RMSE by 0.004 and the MAE by 0.003 on average. In addition, ablation experiments show that the addition of the multi-encoder structure and the BHTP module improves the RMSE and MAE by 0.008 and 0.007 on average, respectively. These results highlight the robustness and utility of the proposed method in real battery management systems, especially under data-scarce conditions.

Keywords: lithium-ion battery; small-sample data; battery state of health; deep learning

1. Introduction

Battery management systems play a critical role in ensuring the safe and stable operation of lithium batteries and support battery recycling for secondary use. This, in turn, helps to reduce greenhouse gas emissions, mitigate environmental pollution, address the energy crisis, and achieve a carbon peak and a carbon-neutral society. The state of health (SOH) of a battery, as one of the parameters of a battery management system, is a crucial indicator of the continued functioning of the battery. Severe degradation of the SOH of a lithium battery can lead to frequent failures [1–3]. Hence, research on SOH can help to identify internal hazards and estimate battery lifecycles so as to avoid further safety issues and facilitate the recycling and reuse of batteries [3–7].

Currently, SOH prediction methods can be categorized into two types: traditional model-based methods, including equivalent circuit models (ECM) [8], electrochemical models [9], and empirical models [10], and data-driven methods. Messing et al. [11,12] demonstrated that electrochemical impedance spectroscopy (EIS) can effectively track the short-term relaxation effects of batteries at different SOHs, and proposed an SOH estimation method combining fractional-order impedance modeling and short-term relaxation effects with EIS characteristics. Luo et al. [13] proposed an EIS-based rapid SOH prediction method by constructing a mathematical model to estimate the SOH of decommissioned batteries. This model combines the constant phase element parameters of EIS-ECM with the state of charge (SOC) and SOH of batteries. Bartlett et al. [14] combined an electrochemical model with an extended Kalman filter to estimate the SOH of lithium-ion batteries with composite electrodes, using a reduced-order electrochemical model to estimate the loss of active materials in batteries to facilitate SOH prediction. However, these methods require specified experimental conditions and are therefore more suitable for battery design and technical improvement.

Among the data-driven methods, Yang et al. [15,16] used four characteristic parameters extracted from the charging curve as the input to a Gaussian process regression model, and then used gray correlation analysis to assess the correlation between the four characteristic parameters and SOH, thus improving the covariance function design and similarity measure to enhance the accuracy of SOH estimation. Tan et al. [17,18] proposed an SOH prediction method based on transfer learning [19]. Zhang et al. [20,21] established a transfer learning model and used the fine-tuning model to conduct SOH estimation by using a small amount of recent historical data of the target battery [22]. Cheng et al. [23] proposed an SOH prediction method that combines empirical mode decomposition with a backpropagation long short-term memory (B-LSTM) neural network [24]. Chang et al. [25,26] proposed an SOH estimation method based on a wavelet neural network combined with a genetic algorithm (GA). After obtaining incremental capacity curves, significant health characteristic variables were extracted using the Pearson correlation coefficient, and the initial parameters of the wavelet neural network were optimized using the GA. Tian et al. [27] proposed a lithium-ion battery SOH estimation method using DT-IC-V health features extracted from a partial charging segment, a DBN-ELM model optimized by SSA, and verified its high accuracy with low errors on Oxford and NASA datasets. Li et al. [28] proposed a SOH prediction method based on neural basis expansion analysis (NBEATS) neural network of hunter and prey optimization (HPO) with variational mode decomposition (VMD). Shi's [29] SOH prediction of lithium-ion batteries uses automatically extracted features and an IWOA-SVR model.

Due to the highly complex internal mechanisms of batteries and the intricate electrochemical reactions involved in their functioning, it is challenging to directly establish a traditional model for battery SOH prediction based on electrochemical principles [30]. In recent years, data-driven methods have been widely and successfully applied for predicting the SOH of batteries due to their high feasibility [31]. However, data-driven models require extensive training, and their accuracy heavily depends on the quality and quantity of the data used for modeling. Factors such as inconsistent usage scenarios and the long lifecycles of batteries often result in limited acquired data and an imbalanced data distribution. Under these conditions, data-driven models often suffer from low accuracy and generalizability. Under small-sample conditions, the differences between battery cells and the capacity recovery phenomenon will lead to unreasonable SOH degradation, making it difficult for a deep learning model to capture the SOH cycle features of batteries [32].

To tackle the challenges of state-of-health prediction for lithium batteries under small-sample conditions, we propose a multi-encoder and transfer learning empowered method, which leverages a multi-encoder structure. From the perspective of the manifold learning theory, data tend to be distributed on low-dimensional manifold structures in a high-dimensional space, and the manifold structures of different sample sets are essentially similar [33]. Based on this theory, a multi-encoder structure is designed, which includes token, position, and time encoders to comprehensively extract multi-dimensional features from the state of health (SOH) sequences. These features reflect the key information of the data on the manifold structure. The BHTP (battery health temporal prediction) module is integrated, and advanced technologies such as residual connections and hierarchical normalization are utilized to enhance feature fusion and stability. In the pre-training phase of transfer learning [34], the model captures the general characteristics of the manifold structure in large-scale data. When fine-tuning, the similarity of the manifold structures of the large and small sample sets is exploited to enable the model to quickly adapt to small-sample tasks [35]. This effectively solves the problem of overfitting, improves the generalization ability of the model, and thus achieves accurate prediction of the SOH of lithium batteries under the condition of scarce data.

The main contributions of this paper are summarized as follows:

(1) We propose a novel multi-encoder framework for SOH prediction of lithium-ion batteries, which integrates element, positional, and temporal encoders to extract multidimensional features, enabling comprehensive and accurate analysis of SOH sequences under small-sample conditions.
(2) We develop an innovative BHTP module for feature fusion and encoding optimization, incorporating residual connections, convolutional layers, and layer normalization, which significantly enhances the model's representational capability, mitigates overfitting risks, and improves stability and generalization.
(3) We propose a transfer learning strategy that combines pretraining with fine-tuning, optimizing model parameters for small-sample scenarios, significantly improving prediction accuracy while reducing computational complexity.
(4) The experiments on NASA datasets show that the proposed method excels under small-sample conditions, achieving fast convergence, significant reductions in RMSE and MAE, and accurate SOH predictions, demonstrating its effectiveness and practicality.

This work proceeds as follows: Section 2 reviews related work. Section 3 provides a detailed explanation of the proposed prediction framework, including the multi-encoder structure, BHTP module, and the transfer learning strategy combining pre-training and fine-tuning. Section 4 presents experimental evaluations, including convergence tests, comparative experiments, and ablation studies. Finally, Section 5 concludes the study and discusses future research directions.

2. Related Works

A small sample size refers to a limited number of cases, rather than a small absolute number of cases. Generally, when a sample size (n) is less than 50, it is considered a small sample [36]. In predicting the SOH of lithium batteries under small-sample conditions, existing methods to mitigate challenges brought by small sample sizes include data augmentation and the integration of prior knowledge [37].

Data augmentation: The core idea of data augmentation is to acquire prior knowledge for augmenting the sample size, thereby expanding the sample data [30]. Li et al. [38,39] introduced the concept of mega-trend diffusion (MTD). This method first identifies the sample boundaries and then augments the sample according to the exhibited trend, ultimately augmenting the sample size. Kang et al. [40] improved the original MTD algorithm to ensure that the range of virtual samples is kept under control [41]. Aiming at the problem of insufficient degradation data of lithium-ion battery, a generalized trend diffusion virtual sample generation method based on differential evolution was proposed [42]. Cheng et al. [43] proposed a lithium-ion battery state estimation model based on data augmentation, which augments the existing dataset by generating synthetic data. Experimental results show that the proposed method can achieve accurate prediction under the condition of a small sample size [44]. Liu et al. [45] augmented the existing dataset by training the collected historical charging voltage data with CGAN to generate new data. Experimental results show that the proposed method significantly improves the estimation performance of lithium battery SOH in the case of different degrees of data loss, and shows a good prediction effect under the condition of a small sample size. Ren et al. proposed to utilize CrGAN for data augmentation of all cells in a battery pack [46]. The model fuses shape and time-dependent features by introducing a cross-attention mechanism and delayed spatial encoding. Guo et al. proposed a lithium battery SOH estimation method based on sample data generation and a temporal convolutional neural network for a data-driven model in the case of limited data [47]. The variational autoencoder (VAE) was used to learn the characteristics and distribution of sample data to generate highly similar data and enrich the number of samples.

MTD is based on the sample trend expansion, focusing on the overall distribution. CGAN and other deep learning, or conditional, or re-association. However, data augmentation methods often produce synthetic samples with significant randomness and poor quality, which deviate from the true data distribution [48]. This deviation can adversely impact the model's predictive accuracy, limiting its reliability and effectiveness [49].

Integrated prior knowledge: Thompson et al. [50] proposed a modeling method combining prior knowledge with neural networks, where prior knowledge is introduced into the neural network with constraints. The prediction results of the constructed network model are credible and consistent with the actual aging process of lithium batteries. Zhao et al. [51] addressed transformer fault issues using a residual back propagation neural network, combining multiple residual modules [52] to ensure that prior knowledge aligns with the needs of the network structure and improves the transformer fault prediction accuracy under small-sample conditions. Aiming at the problem of low accuracy of RUL prediction results under the condition of small samples, Hao Keqing [53] proposed to obtain prior knowledge by drawing the curve relationship between the extracted indirect health factors and the lithium battery capacity. The RUL prediction model [54] of lithium battery with fusion constraints was established. Methods involving the integration of prior knowledge not only adhere to the network convergence conditions but also make full use of existing knowledge. Numerous experiments have achieved improved results when using such methods. Jiang et al. [55] proposed a battery SOH estimation method combining EIS feature extraction and Gaussian process regression. The feature selection was integrated into the prior knowledge of battery electrochemical characteristics, and the model performed well under fixed frequency impedance characteristics. Li et al. [56] proposed a lithium battery internal temperature estimation method based on multi-frequency virtual part impedance and a GPR model. By using an equivalent circuit model, sensitivity analysis, and Pearson coefficient screening features, the prior knowledge of the electrochemical process was introduced.

Different methods for integrating prior knowledge have obvious differences. In terms of the way of introducing knowledge, Thompson et al. introduced prior knowledge into the neural network through constraints, while Zhao et al. made the prior knowledge adapt to the network structure by means of multiple residual modules. Regarding the acquisition of knowledge, Hao Keqing obtained prior knowledge by constructing the curve relationship between the extracted indirect health factors and the battery capacity.

Jiang et al. integrated feature selection into the knowledge of battery electrochemical characteristics, and Li et al. introduced the knowledge of the electrochemical process by using the equivalent circuit model and other methods. However, the integration of prior knowledge in the aforementioned methods often depends on domain-specific heuristic rules, limiting their ability to dynamically model complex data relationships [57]. To address these challenges, an innovative approach is proposed, combining a multi-encoder structure, the BHTP module, and a transfer learning strategy. This design significantly improves the accuracy, stability, and generalization capability of lithium battery SOH prediction, effectively addressing the shortcomings of data augmentation methods and prior knowledge-based techniques.

3. Multi-Encoder Architecture and BHTP-Enhanced Feature Fusion

3.1. Task Definitions

The SOH of a battery is defined as the ratio of its current capacity to its rated capacity [58]. The full lifecycle SOH sequence of a battery is denoted as: SOH = {h1, h2, . . ., hn}, where hn is the SOH value corresponding to the n-th cycle. Based on the top segment of the battery's SOH sequence, SOH_A = {h1, h2, . . ., hm | 1 < m < n}, the SOH of the battery is predicted, and the full lifecycle SOH sequence is obtained as follows: SOH_B = {hm + 1, hm + 2, . . ., hn}.

3.2. Sliding Window Prediction

In this study, we used the sliding window technique to predict the SOH sequence of a lithium battery [59]. This technique is used for prediction on temporal datasets. A fixed-size window is applied to the top segment of the SOH sequence, covering the SOH sequence from the 1st to the m-th cycle. The SOH value for the m + 1-th cycle is predicted using the SOH sequence from the 1st to the m-th cycle. The window then slides forward by one step to cover the SOH sequence from the 2nd to the m + 1-th cycle to predict the SOH value for the m + 2-th cycle. This process is repeated iteratively until the SOH value for the n-th cycle is predicted.

3.3. Multi-Encoder Architecture: Overall Structural Design

In this study, we used an autoencoder model to predict the SOH of lithium batteries under small-sample conditions. The overall model design is shown in Figure 1. An autoencoder consists of two components: an encoder and a decoder. It captures data features through the encoding and decoding processes and then outputs the results. Based on the construction of an autoencoder, this study introduced a multi-encoder structure and incorporated a BHTP module. The multi-encoder structure extracts the multi-perspective features of a lithium battery SOH sequence, while the BHTP module integrates and processes the encoded features to generate the codes for the predicted SOH values of the lithium battery.

To prevent model overfitting under small-sample conditions, a combined pre-training and fine-tuning strategy was used to predict the SOH values of lithium batteries. In the pre-training phase, the input data comprises a large dataset of full lifecycle SOH sequences from lithium batteries. In order to ensure the stability of the training model, the data are first normalized by the following formula, where μ and σ are the mean and RMSD of the series, respectively:

$$h_i^{norm} = \frac{h_i - \mu}{\sigma}, i = 1, 2, \ldots, n \quad (1)$$

where h_i^{norm} represents the normalized value of h_i, h_i denotes the i-th element in the original data sequence, μ stands for the mean value of the data, and σ represents the standard deviation of the data, with $i = 1, 2, \cdots$, n. This formula implies that it normalizes the original data element h_i by subtracting the mean μ and then dividing by the standard deviation σ.

$$\mu = \frac{1}{N}\sum_{i=1}^{N} h_i \tag{2}$$

where μ represents the mean value of the data, N is the total number of data elements, and h_i stands for the i-th data element, with i ranging from 1 to N. This formula shows that the mean μ is calculated by summing up all N data elements h_i and then dividing by the total number N.

$$\sigma = \sqrt{\frac{1}{N}\sum_{i=1}^{N}(h_i - \mu)^2} \tag{3}$$

where σ represents the standard deviation of the data, N is the total number of data elements, h_i denotes the i-th data element, and μ is the mean value of the data, with i ranging from 1 to N. This formula indicates that the standard deviation σ is obtained by first computing the square of the difference between each data element h_i and the mean μ, summing these squared differences, dividing by the total number N, and finally taking the square root.

The consistency and reliability of processing different battery SOH sequence data sets are improved by instance normalization. The data are then divided by a sliding window:

$$Sliding\ Window_k = \{h_k, h_{k+1}, \ldots, h_{k+m-1}\}, k = 1, 2, \ldots, N \tag{4}$$

where $SlidingWindowk$ represents the k-th sliding window; h_k, h_{k+1}, $\cdots$, h_{k+m-1} denote the elements in the data sequence; and k = 1,2, $\cdots$, N. This formula indicates that the k-th sliding window consists of m consecutive elements starting from the k-th element in the data sequence.

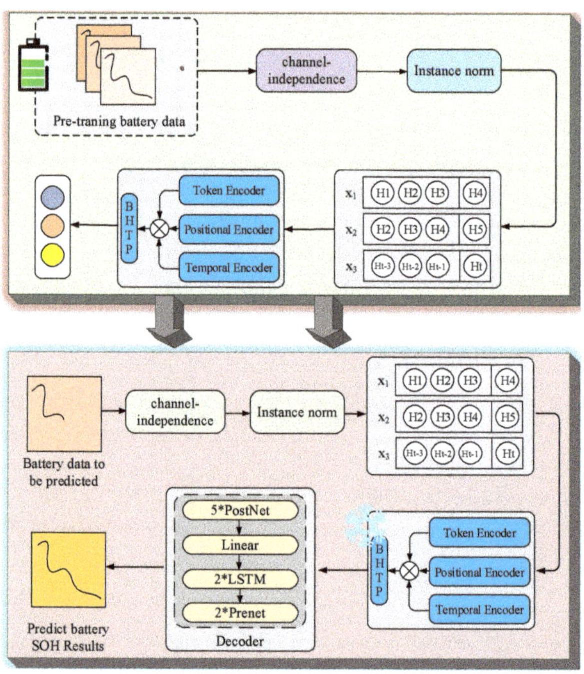

Figure 1. Autoencoder-based SOH prediction method for lithium batteries.

Finally, the model is trained using the multi-encoder structure and the BHTP module. After training, the BHTP module is frozen, and the weights are transferred to the fine-tuning phase.

In the fine-tuning phase, the input data comprises the small-sample SOH sequence of the lithium battery being used for prediction. After undergoing instance normalization and channel independence, the data are encoded by the multi-encoder structure and output to the BHTP module. The decoder then decodes the output to generate the predicted SOH values for the time instance. Since the BHTP module is frozen during the pre-training phase, its parameters remain unchanged during the model training phase, and only the encoding and decoding parts are updated. This approach not only leverages the feature representation ability learned by the BHTP module during pre-training but also allows other parts of the model to make adaptive adjustments, thereby achieving more efficient and accurate prediction of the state of health (SOH) of lithium-ion batteries under small-sample conditions. Figure 2 presents the workflow of the proposed battery SOH prediction method under small-sample conditions.

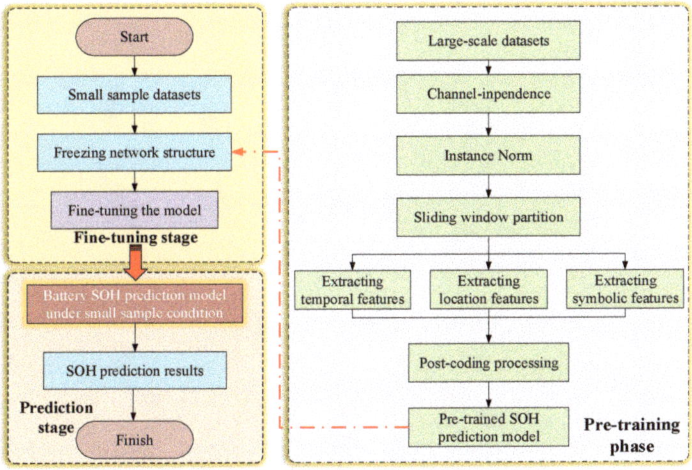

Figure 2. Workflow design for the proposed prediction approach.

3.4. Advanced Multi-Encoder Architecture

The prediction of a lithium battery's SOH involves multi-dimensional characteristic information, including the battery's chemical properties, physical state, usage history, environmental conditions during usage, and its temporal degradation trend. A single type of encoder is unable to capture such comprehensive, multi-dimensional, and multi-perspective data, and therefore will limit the accuracy and generalizability of the prediction model [60,61]. In this study, we used a multi-encoder structure—comprising a token encoder, a positional encoder, and a temporal encoder—to comprehensively analyze and represent the features in the SOH data of lithium batteries. The multi-encoder structure is shown in Figure 3. Through the coordinated processes of these three encoders, the multi-encoder structure proposed in this study extracts the multi-dimensional features from the SOH data to provide feature codes for the subsequent small-sample SOH prediction model for lithium batteries, thus enhancing the accuracy and generalizability of the model. The specific formula is as follows, where Z is the feature information and J is the encoder number:

$$Z_k^j = Encoder(\text{Sliding Window}_\text{k}), \ \ j = 1,2,3 \tag{5}$$

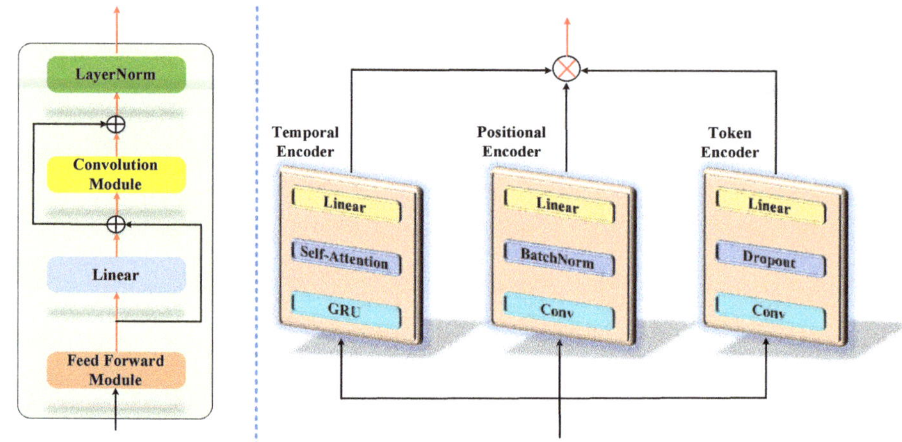

(a) Structure of the BHTP module

(b) Structure of the Multi-encode

Figure 3. Multi-encoder structure and BHTP module structure diagram.

Token encoder: The token encoder is a key component of data preprocessing, which is responsible for adapting the dimensions of the input data to the requirements of the subsequent processing module. Due to the diversity and heterogeneity of the original data features, the feature scales are different. Through the transformation and mapping operations such as standardization, normalization, and coding, the various features are unified into a format and scale that can be recognized by the model to ensure that the data can be effectively processed by the model.

$$Z = XW + b \tag{6}$$

where X represents the input data, W denotes the weight matrix, and b represents the bias vector. This formula indicates that the linear layer maps the input features to new features by performing a linear transformation on the input data.

$$Z_{drop} = \frac{M \odot Z}{1 - p} \tag{7}$$

where Z represents the output of the linear layer, M denotes a mask matrix with elements determined by the dropout probability p, and $1 - p$ is the probability of an element being kept. This formula indicates that during training, dropout randomly sets some values in Z to 0 to prevent overfitting.

$$Y_{\text{token}} = \text{Conv}\left(Z_{\text{drop}}, K\right) + b_{\text{conv}} \tag{8}$$

where Z_{drop} represents the output of the dropout layer, K denotes the convolutional kernel, and b_{conv} represents the bias of the convolutional layer. This formula indicates that the convolutional layer extracts local features from the input data by convolving Z_{drop} with K and adding the bias.

Positional encoder: The positional encoder is used to capture the relative position information between elements in the sequence data. In the time series data type, the order of the elements carries important semantics. It assigns each element a unique vector representation related to its position. The vector contains information about the absolute position of the element in the sequence, and can also imply the relative position relationship by encoding. It can significantly improve the understanding ability and prediction accuracy of the BHTP model for sequence data.

$$Z_{\text{pos}} = XW_{\text{pos}} + b_{pos} \tag{9}$$

where X represents the input data, W_{pos} denotes the weight matrix, and b_{pos} represents the bias vector. This formula indicates that the linear layer in the positional encoder transforms the input features to prepare for subsequent operations.

$$z\text{pos}_{\ norm_{ijt}} = \gamma_j \hat{z}pos_{ijt} + \beta_j \tag{10}$$

where $\text{z}pos_{norm_{ijt}}$ represents the transformed element in the positional encoder's BatchNorm layer, $\hat{\text{z}}pos_{ijt}$ denotes the normalized element, γ_j represents the learnable scaling factor for feature channel j, and β_j denotes the learnable offset factor for feature channel j. This formula indicates that the normalized element $\hat{\text{z}}pos_{ijt}$ is further adjusted by the learnable factors γ_j and β_j to help the model's performance.

$$Y_{pos} = \text{Conv}\left(Z_{pos_{norm}}, K_{pos}\right) + b_{pos_{\text{conv}}} \tag{11}$$

where $Z_{pos_{norm}}$ represents the output of the batch normalization layer, K_{pos} denotes the convolutional kernel, and $b_{pos_{\text{conv}}}$ represents the bias of the convolutional layer. This formula indicates that the convolutional layer in the positional encoder extracts position-related features from the normalized input data.

Temporal encoder: The state of health (SOH) of a lithium battery has a timing feature that reflects the changing trend of the battery's performance over time. Affected by the number of charging and discharging, environmental temperature, and other factors, the battery performance changes dynamically. The time series encoder extracts and represents this time series information by processing the input sequence step by step, learns the time series dependence and dynamic change pattern, and mines the key information in the lithium battery SOH data.

$$Z_{temp} = XW_{temp} + b_{temp} \tag{12}$$

where X represents the input data, W_{temp} denotes the weight matrix, and b_{temp} represents the bias vector. This formula indicates that the linear layer in the temporal encoder maps the input features to a new space to prepare for processing time-series data.

$$\text{Self} - \text{attention}\ (Q, K, V) = \text{softmax}\left(\frac{QK^T}{\sqrt{d_{\text{att}}}}\right)V \tag{13}$$

where Q, K, and V are matrices obtained by linearly transforming the input data, and d_{att} denotes the attention dimension. This formula indicates that the self-attention mechanism captures long-range dependencies in the time-series data by calculating attention scores, normalizing them, and performing a weighted sum on the value matrix.

$$h_t = (1 - f_z(x_t, h_{t-1})) \odot f_j(f_r(x_t, h_{t-1}), x_t, h_{t-1}) + f_z(x_t, h_{t-1}) \odot h_{t-1} \tag{14}$$

where x_t represents the input at time step t, $ht - 1$ denotes the hidden state at the previous time step, f_r calculates the reset gate, f_z calculates the update gate, and $f_{\tilde{h}}$ calculates the candidate hidden state. This formula indicates that the GRU layer determines how to combine the previous hidden state and the candidate hidden state to form the current hidden state, capturing temporal dependencies in the sequence data.

$$Y = Y_{\text{token}} \odot Y_{\text{pos}} \odot Y_{\text{temp}} \tag{15}$$

where Y_{token}, Y_{pos}, and Y_{temp} represent the outputs of the token encoder, positional encoder, and temporal encoder, respectively. This formula indicates that the final output is obtained by multiplying element-wise the outputs of the three encoders to combine the features they extracted.

3.5. Residual-Fused BHTP Module

To process the SOH codes of a lithium battery, a BHTP module is used to accurately predict SOH under small-sample conditions. The BHTP module is used to process encoded SOH data by first using an encoding layer to pre-process and encode the raw SOH data of the lithium battery. The resulting characteristic data are then passed to the core of the BHTP module for further processing. The core of the BHTP module comprises key modules combined through residual connections. These modules include the FeedForward module, linear layer, convolution module, and LayerNorm module. The specific BHTP model is shown in Figure 3. The residual connections mitigate the vanishing gradient issue during the deep network training, maintaining the stability and efficiency of network learning.

First, the input data dimensions are changed to pass into the FeedForward module, which maps its dimensions into a higher-dimensional space suitable for subsequent processing. The assisting model of the rectified linear unit activation function captures and learns the nonlinear features of the small-sample data during training. The convolution module enables the model to learn broader contextual information about the data. In the convolutional layers, the convolution kernel slides with the local element-by-element multiplication of the input data to extract local features such as edges. Multiple convolution layers are concatenated, and the later layer takes the output of the previous layer as the input to expand the coverage of features. The LayerNorm module normalizes the mean and variance of the inputs for each layer, thus reducing the impact of internal covariate shifts. The final linear module obtains the feature representations output from the FeedForward module through linear transformation. After being processed by the BHTP module, the data are transformed into highly abstract and information-rich feature representations. The formula is as follows:

$$Z_k^{fuse} = BHTP(Z_k^1, Z_k^2, Z_k^3) \tag{16}$$

where Z_k^{fuse} signifies the fused feature representation. These feature representations are fed into the decoder part for the final prediction task. According to the relationship between the feature representation and the predicted target, the decoder outputs the predicted value of the SOH of the lithium battery.

3.6. Adaptive and Hybrid Transfer Learning

To better deal with the small sample challenge in a lithium battery's health state (SOH) prediction, in the above-proposed method, we improved the performance of lithium battery SOH prediction under small sample conditions by the transfer learning method.

Transfer learning is a machine learning method that acquires knowledge in a source task and applies it to a target task. It is particularly suitable for small sample data. It improves the generalization ability of the model by pre-training it on a large dataset and fine-tuning it on a new task. The core idea of transfer learning is that although there are differences between the source and target domains, their underlying feature spaces have certain structural similarities. Therefore, learning the features of large sample data and migrating them to small sample data can effectively learn prior knowledge and improve the prediction ability of the model under small-sample data. The transfer learning model design is shown in Figure 4. In this study, the low-dimensional manifold features learned in large-scale SOH sequences using a multi-encoder structure are transferred by freezing the BHTP module to quickly adapt to the target small-sample task, thereby improving the prediction accuracy of the model.

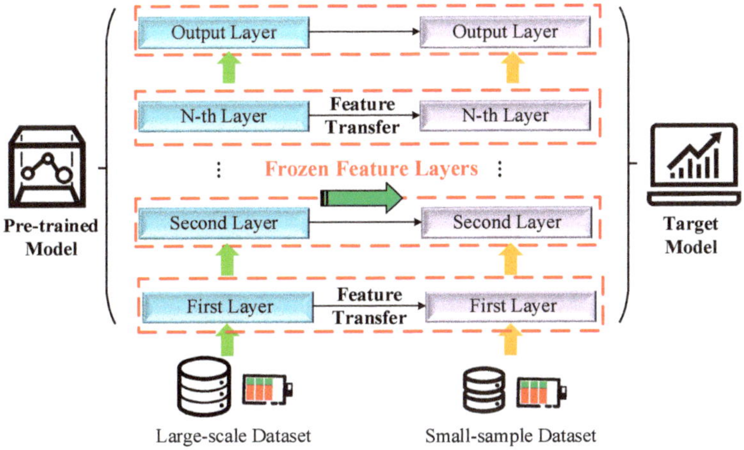

Figure 4. Adaptive and hybrid transfer learning design.

Firstly, in the pre-training stage containing a large number of lithium battery life cycle SOH sequences, the model preprocesses the data by instance normalization and sliding window division, uses the multi-encoder structure to extract features from multiple perspectives, and performs feature fusion through the BHTP module to obtain a common feature representation. Subsequently, in the fine-tuning phase, the model parameters obtained by pre-training are transferred to the target model, and the pre-trained parameters of the BHTP module are frozen. The formula is as follows:

$$\theta_{\text{init}} = \theta_{pre} \tag{17}$$

where θ_{init} represents the initial parameters used in the fine-tuning phase, while θ_{pre} denotes the parameters obtained during the pre-training phase. This formula indicates that the parameters learned in the pre-training stage are directly transferred as the initialization parameters for the fine-tuning phase to ensure knowledge reuse.

$$\theta_{BHTP}^{fine-tune} = \theta_{BHTP}^{pre} \tag{18}$$

where $\theta_{BHTP}^{fine-tune}$ refers to the parameters of the BHTP module during the fine-tuning stage, and θ_{BHTP}^{pre} represents the parameters of the BHTP module trained in the pre-training stage. The encoder, after parameter transfer, is used to extract features to adapt to the prediction requirements of small samples, so as to generate the prediction value through the decoder. The formula is as follows:

$$\hat{h}_{k+m} = Decoder(Z_k^{fuse}) \tag{19}$$

where $\hat{h}_{k+m}$ denotes the predicted value, specifically the state of health (SOH) of the lithium batteries. By minimizing the fine-tuning loss function, the weights of the multiple encoders and decoders in the model are optimized to better adapt to the characteristics of small-sample data while keeping the parameters of the BHTP module unchanged.

$$L_{finetune} = \frac{1}{N'} \sum_{k=1}^{N'} \left(h_{k+m} - \hat{h}_{k+m}\right)^2 \tag{20}$$

where h_{k+m} represents the actual SOH value of the battery, while $\hat{h}_{k+m}$ corresponds to the predicted SOH value. N' is the total number of samples in the dataset. This process not only effectively alleviates the overfitting problem caused by small samples, but also enhances the generalization ability of the model, so as to conduct the high-accuracy lithium battery SOH prediction under the condition of limited data.

4. Experiment and Results Analysis

4.1. Dataset and Experimental Setup

The dataset used in this study is the battery cycle life dataset collected from lithium cobalt oxide ($LiCoO_2$) 18650 batteries by the National Aeronautics and Space Administration (NASA) [62]. The data in Table 1 indicate that all batteries were rated 2 Ahr. The aging simulation of a lithium battery involves two processes: charging and discharging. During the charging process, the battery was charged with a constant current (1.5 A) until its voltage reached 4.2 V. Then, charging continued in a constant voltage mode until the current dropped to 20 mA. During the discharge process, the discharge was carried out at a constant current of 2 A until the voltage of B0005 dropped to 2.7 V and the voltage of B0007 dropped to 2.2 V. For B0033, the discharge was carried out at 4 A until its voltage dropped to 2.0 V. The experimental termination condition for batteries B0005 and B0007 is that the rated capacity is reduced to 1.4 Ahr, with a 30% decrease. The experimental termination condition for battery B0033 is that the rated capacity is reduced to 1.6 Ahr, with a 20% attenuation. Different pre-set values were used to simulate different aging degrees.

Table 1. Introduction to the battery dataset.

Dataset	Temperature	Condition of Charge	Condition of Release	Rated Capacity	End Condition
B0005	24 °C	1.5 A:4.2 V → 20 mA	2 A → 2.7 V	2 Ahr	1.4 Ahr
B0007	24 °C	1.5 A:4.2 V → 20 mA	2 A → 2.2 V	2 Ahr	1.4 Ahr
B0033	24 °C	1.5 A:4.2 V → 20 mA	4 A → 2.0 V	2 Ahr	1.6 Ahr

In this study, we selected three individual battery cells (B0005, B0007, and B0033) from the dataset for analysis. The charging and discharging experiments were conducted on all three batteries at a room temperature of 24 °C. Since batteries B0005 and B0007 have the same discharge current and contain historical data of 168 charge/discharge cycles, these two cells are placed in the same battery data group. Since batteries B0005 and B0033 have different discharge currents and battery B0033 contains historical data of 197 charge/discharge cycles, B0005 and B0033 are placed in different battery data groups. Due to differences in charge/discharge strategies, the SOH curves of the lithium batteries differ significantly, as shown in Figure 5.

From Figure 5, the following observations can be made: Firstly, it can be seen from the three figures that with the increase of battery operation cycle times, side reactions between electrodes and electrolytes occur, resulting in continuous loss of lithium and active substances, making the SOH of the three batteries show an overall decline trend. Secondly, the overall trend of the battery SOH curve in the figure is decreasing, but it is not monotonically decreasing, and there is an intermittent rise. Since the battery's SOC, ambient temperature, and circulating current can significantly affect its degradation process, it is difficult to directly measure the theoretical SOH value of the battery. This is shown in Figure 5 as an intermittent rise in the battery's SOH curve. Finally, the SOH curves of the batteries from the same group (B0005 and B0007) differ only slightly from each other, while those of batteries from different groups (B0005 and B0033) differ significantly from each other.

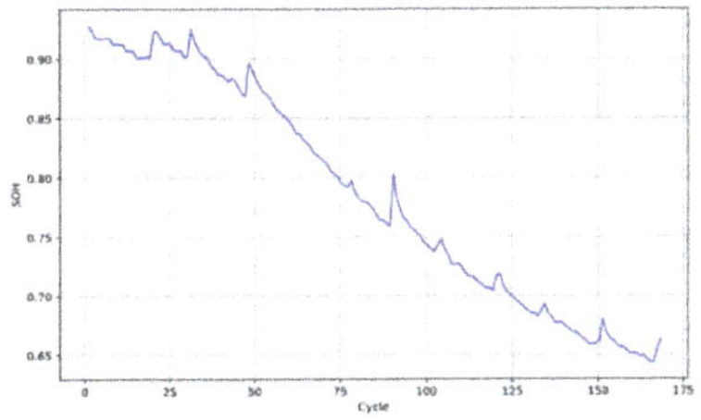

(a)

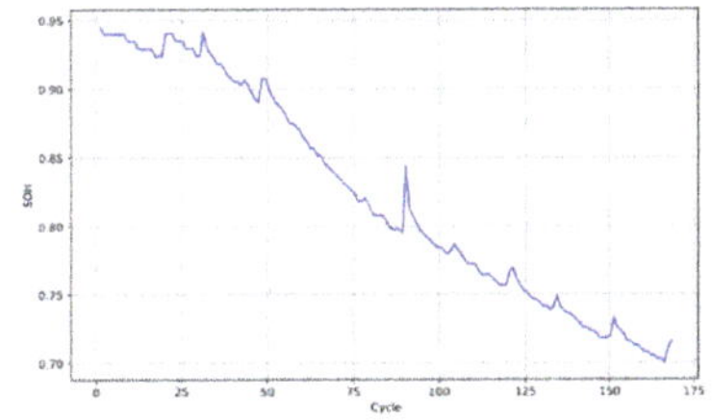

(b)

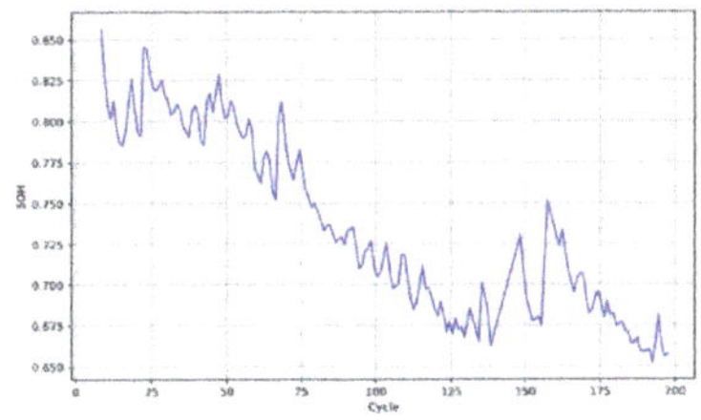

(c)

Figure 5. SOH decline graphs for (**a**) battery B0005, (**b**) battery B0007, and (**c**) battery B0033.

Table 1 shows that different battery packs have different experimental conditions and different termination conditions, which can be obtained as follows: There is a high degree of similarity in the data of batteries from the same group, and a low degree for batteries in different groups. Therefore, to investigate the results of the proposed algorithm in predicting the SOH values of lithium batteries when the data similarities in the pre-training and the fine-tuning phases differ, the complete B0005 dataset was used for pre-training. For fine-tuning, the SOH values corresponding to the first 20% (i.e., the SOH values within the initial 20% of the battery SOH cycle sequences), the first 30% (i.e., the SOH values within the initial 30% of the battery SOH cycle sequences), and the first 40% (i.e., the SOH values within the initial 40% of the battery SOH cycle sequences) of the battery SOH cycle sequences in the B0007 and B0033 datasets were used. The remaining portions of these datasets served as the test sets. The selection of data for fine-tuning is performed to verify the adaptability of different-sized datasets during the transfer learning process. Smaller-scale datasets (such as the first 20% of the data) can better simulate small-sample scenarios and are used to test the model's ability to make accurate predictions using pre-trained knowledge with limited information. Medium-scale (the first 30% of the data) and relatively larger-scale (the first 40% of the data) datasets can demonstrate the model's adaptability under different data richness conditions. The Adam optimizer was used for training with an initial learning rate of 0.003, a batch size of 128, and a number of epochs of 500. The sliding window size is 7.

4.2. Evaluation Indicators

To measure the differences between the predicted and true values and evaluate the prediction results, root mean squared error (RMSE) and mean absolute error (MAE) were used as evaluation indicators of the algorithm's performance. RMSE and MAE indicate the error between predicted and true values. The lower the RMSE and MAE values, the closer the predicted value is to the true value, and the better the prediction result. RMSE and MAE may obtain values in the range $[0, +\infty)$, expressed as follows:

$$RMSE = \sqrt{\frac{\sum_{i=1}^{n} (\hat{y}_i - y_i)^2}{n}} \quad (21)$$

$$MAE = \frac{\sum_{i=1}^{n} |\hat{y}_i - y_i|}{n} \quad (22)$$

In Equations (1) and (2), n is the total number of samples; $\hat{y}_i$ is the predicted value; and y_i is the true value.

4.3. Convergence Experiment

Using the proposed method, training was conducted on the top 30% of the B0007 and B0033 datasets. The convergence results are shown in Figure 6. To accurately record the model training curves, both groups of batteries underwent 500 epochs of training. After training, Figure 6 shows that the model achieved convergence on both datasets and was able to reach the convergence values rapidly. Comparing the two curves, it can be observed that on the B0007 dataset, the training and test curves exhibited a more stable overall performance, with relatively smaller initial fluctuations and a smaller gap between the two curves in the later stages. Both curves ultimately converged to lower values, indicating good convergence performance and a lower risk of overfitting. In contrast, for the B0033 dataset, the training curve exhibited larger initial fluctuations and continued to show noticeable minor fluctuations throughout the training process. Similarly, the test curve also had significant initial fluctuations, and in the later training stages, the gap between the training and test curves was slightly larger compared to the B0007 dataset. This suggests that the model's convergence performance was relatively poorer on the B0033 dataset. The better convergence observed on the B0007 dataset can be attributed to its greater similarity to the B0005 dataset compared to the B0033 dataset.

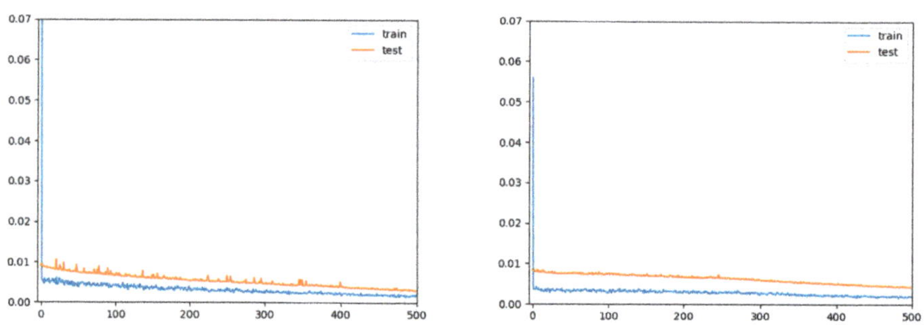

(a) (b)

Figure 6. Convergence experimental results for (**a**) the 30% training set for B0007 and (**b**) the 30% training set for B0033.

4.4. Comparative Experiment

To evaluate the ability of the proposed method to predict the SOH values of lithium batteries under small-sample conditions, a comparative experiment was conducted. Using the top 20–40% of the datasets as training sets, we compared the existing mainstream deep learning models for lithium battery SOH predictions, namely, the long short-term memory (LSTM), bidirectional long short-term memory (BiLSTM), gated recurrent unit (GRU), and bidirectional gated recurrent unit (BiGRU) models.

LSTM [63] model: the LSTM model is essentially a special form of recurrent neural network (RNN) designed to address the short-term memory issue present in traditional RNNs by incorporating gates, and enables the RNN to effectively leverage long-term temporal information.

BiLSTM [64] model: the BiLSTM network is an improved LSTM neural network that captures the contextual information in the input sequence by incorporating bidirectionality into the model, thereby enhancing the sequence modeling performance.

GRU [65] model: a GRU is a neural network architecture for sequence modeling, particularly suitable for applications such as natural language processing and time series analysis. It is closely related to LSTM, as both are designed to address the long-term dependency issue. However, compared to LSTM, the GRU model has a simpler structure and fewer parameters.

BiGRU model: BiGRU is an improved RNN architecture and an extension of a GRU. It consists of two GRU models: a forward GRU model that processes the input in the forward direction, and a backward GRU model that processes the input in the reverse direction. The BiGRU model combines a GRU with bidirectionality to capture efficiently the contextual information in the input sequence.

The prediction results using the different methods are shown in Figure 7.

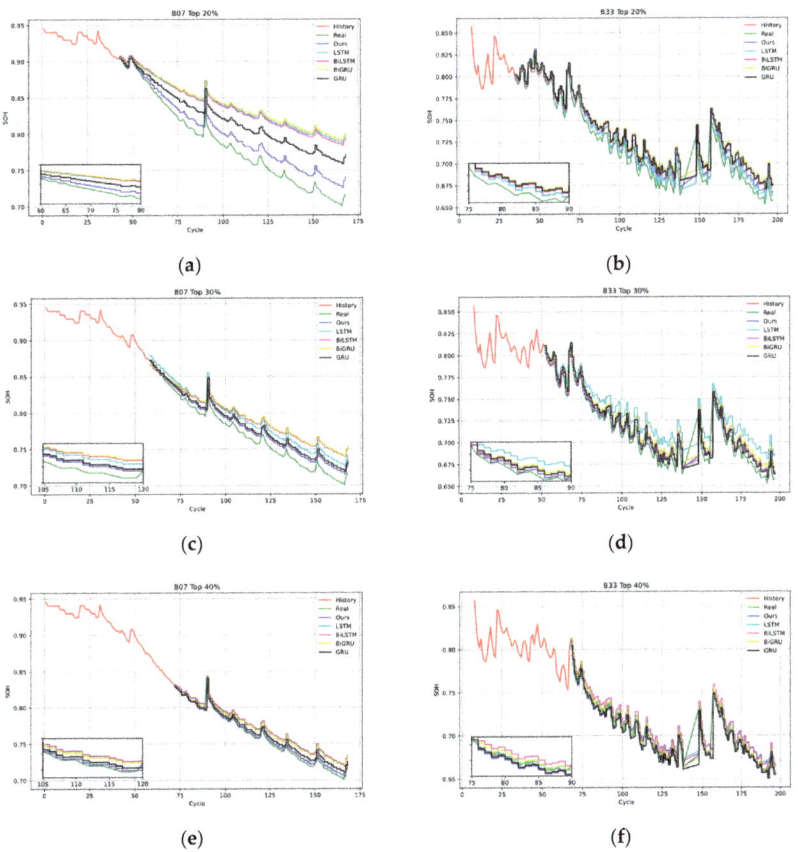

Figure 7. Comparison of experimental results: (**a**) the 20% training set of the B0007 dataset, (**b**) the 20% training set of the B0033 dataset, (**c**) the 30% training set of the B0007 dataset, (**d**) the 30% training set of the B0033 dataset, (**e**) the 40% training set of the B0007 dataset, and (**f**) the 40% training set of the B0033 dataset.

As shown in Figure 7, across the top 20–40% of the B0007 and B0033 datasets, all of the algorithm models exhibit a good fit with the actual SOH curves of batteries. Compared to the baseline deep learning models, the proposed model exhibits an improved fitting effect due to its utilization of prior knowledge obtained from the B0005 dataset. The specific evaluation indicators for each model are listed in Table 2.

Table 2. Results of the comparative experiment.

Dataset	Model	Training Set Division	RMSE	MAE
B0007	Ours	Top20%	0.017	0.015
		Top30%	0.010	0.009
		Top40%	0.002	0.002
	LSTM	Top20%	0.057	0.051
		Top30%	0.020	0.019
		Top40%	0.013	0.012
	BiLSTM	Top20%	0.056	0.051
		Top30%	0.025	0.022
		Top40%	0.013	0.012
	BiGRU	Top20%	0.059	0.053
		Top30%	0.026	0.023
		Top40%	0.011	0.010
	GRU	Top20%	0.038	0.034
		Top30%	0.013	0.012
		Top40%	0.006	0.005
B0033	Ours	Top20%	0.007	0.007
		Top30%	0.004	0.005
		Top40%	0.002	0.003
	LSTM	Top20%	0.007	0.007
		Top30%	0.011	0.010
		Top40%	0.005	0.005
	BiLSTM	Top20%	0.012	0.011
		Top30%	0.007	0.007
		Top40%	0.008	0.008
	BiGRU	Top20%	0.015	0.013
		Top30%	0.010	0.010
		Top40%	0.004	0.003
	GRU	Top20%	0.013	0.012
		Top30%	0.007	0.007
		Top40%	0.004	0.003

The experimental results in Table 2 show that the proposed model achieved significant reductions in RMSE and MAE compared to LSTM, BiLSTM, BiGRU, and GRU across training samples of different sizes. In the B0007 dataset, both RMSE and MAE dropped by an average of 0.020, while in the B0033 dataset, RMSE and MAE dropped by an average of 0.005 and 0.003, respectively. Additionally, more significant improvements were achieved on the 20% training sets than on the 30% and 40% training sets. The prediction results of the proposed model were more accurate and met the specified prediction requirements, thereby proving the effectiveness and superiority of the proposed method in predicting the SOH of lithium batteries under small-sample conditions.

4.5. Ablation Experiment

To verify the effectiveness of each module in the proposed model, an ablation experiment was conducted by using the top 30% of the B0007 and B0033 datasets as the training sets. The experimental results are shown in Table 3. Moreover, the first 20%, 30%, and 40% training sets of the B0007 and B0033 datasets were used for effectiveness experiments, and the experimental results are shown in Figure 8.

Table 3. Results of the ablation experiment.

Dataset	Temporal Encoder	Positional Encoder	Token Encoder	BHTP	RMSE	MAE
B0007	√	√		√	0.012	0.010
	√		√		0.013	0.010
		√	√	√	0.015	0.014
	√	√	√		0.017	0.015
B0033	√	√		√	0.005	0.004
	√		√		0.006	0.005
		√	√	√	0.006	0.006
	√	√	√		0.007	0.007

This study introduced a multi-encoder structure based on an autoencoder and incorporated a BHTP module. The data were output to the decoder to predict the SOH of lithium batteries. The experimental results in Table 3 indicate that the proposed model significantly outperformed the baseline models in terms of the decrease in RMSE and MAE scores. This improvement can be attributed to the multi-encoder structure, which is capable of effectively extracting relevant information from the SOH sequence from multiple perspectives. In addition, the BHTP module greatly improved the prediction accuracy and the performance of the model under small-sample conditions.

From the experimental results in Table 3, when some encoders are retained or all of them are removed, the RMSE and MAE indicators of the model will deteriorate. Taking the B0007 battery data as an example, when all three encoders are retained, the RMSE is 0.012 and the MAE is 0.010. However, after removing some encoders, the RMSE and MAE increase. This indicates that the multi-encoder structure can comprehensively capture multi-dimensional information such as environmental conditions and time-related degradation trends in the battery SOH sequence. The absence of any encoder will cause the model to fail to fully extract features, thereby reducing the prediction accuracy. When the BHTP module is removed, the model's performance significantly declines. In the B0033 battery data, when the BHTP module is included, the RMSE is 0.005 and the MAE is 0.004. After removing this module, the RMSE becomes 0.007 and the MAE becomes 0.007. This shows that the BHTP module effectively fuses the features extracted by the multi-encoder through internal techniques such as residual connections and layer normalization, reduces the problem of gradient vanishing, and improves the prediction accuracy and performance of the model under small-sample conditions. Overall, the multi-encoder structure and the BHTP module cooperate with each other to jointly improve the model performance.

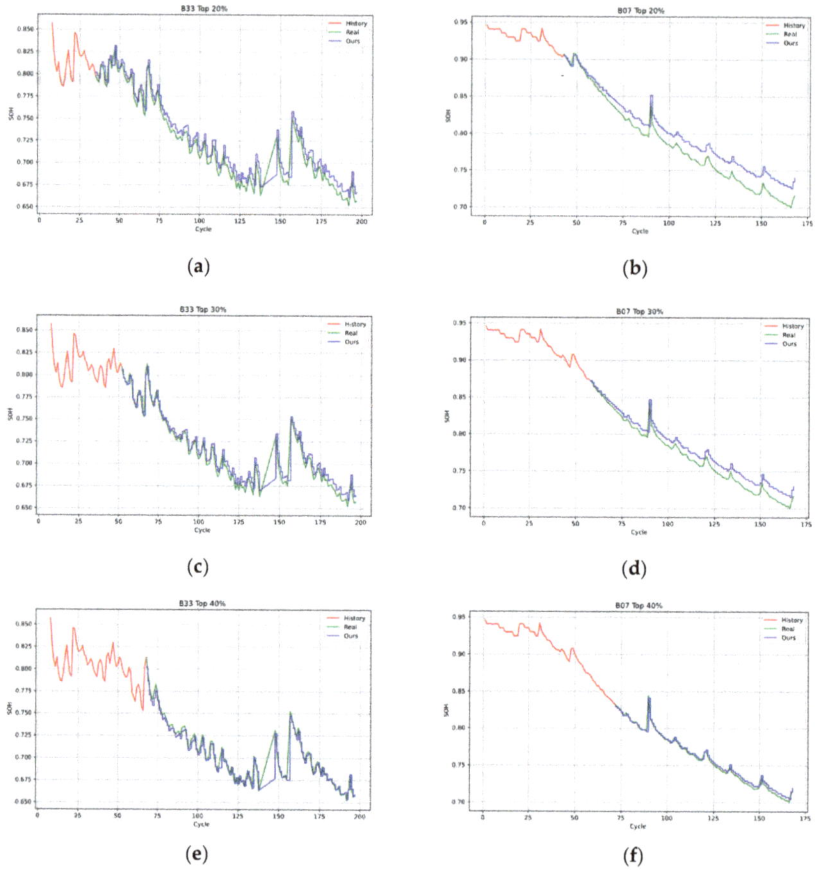

Figure 8. Comparison results show (**a**) the 20% training set of the B0033 dataset, (**b**) the 20% training set of the B0007 dataset, (**c**) the 30% training set of the B0033 dataset, (**d**) the 30% training set of the B0007 dataset, (**e**) the 40% training set of the B0033 dataset, and (**f**) the 40% training set of the B003 dataset.

4.6. Case Study

To evaluate the differences between the predicted and true values, a case study was conducted using the B0007 and B0033 datasets. For each sized training set, four cycle counts were selected to analyze the differences between the predicted and true values. The results are listed in Tables 4–6.

Table 4. Predicted results for the 20% training set.

Dataset	Number of Cycles	Predicted Value	True Value	Difference
B0007	43	0.906	0.905	+0.001
	85	0.814	0.800	+0.014
	127	0.767	0.746	+0.021
	168	0.740	0.716	+0.024
B0033	35	0.802	0.798	+0.004
	89	0.732	0.724	+0.008
	138	0.672	0.663	+0.009
	197	0.667	0.658	+0.009

Table 5. Predicted results for the 30% training set.

Dataset	Number of Cycles	Predicted Value	True Value	Difference
B0007	58	0.872	0.873	−0.001
	95	0.803	0.795	+0.008
	132	0.751	0.738	+0.013
	168	0.730	0.716	+0.014
B0033	52	0.806	0.809	−0.003
	100	0.710	0.705	+0.005
	148	0.734	0.731	+0.003
	197	0.667	0.658	+0.009

Table 6. Predicted results for the 40% training set.

Dataset	Number of Cycles	Predicted Value	True Value	Difference
B0007	72	0.829	0.831	−0.002
	104	0.789	0.787	+0.002
	136	0.741	0.739	+0.002
	168	0.720	0.716	+0.004
B0033	68	0.802	0.812	−0.010
	111	0.700	0.691	+0.009
	154	0.680	0.680	0
	197	0.667	0.658	+0.009

As shown in Tables 4–6, the predicted values are generally higher than the true values, and the errors show an increasing trend as the number of cycles increases. However, the predicted values are still relatively close to the true values, demonstrating the feasibility of SOH prediction under small-sample conditions using this model.

5. Conclusions

Accurate SOH prediction under small-sample conditions is challenging due to individual cell variability, capacity recovery effects, and the difficulty of capturing multi-dimensional degradation features. To address these issues, this study proposes an autoencoder-based framework incorporating a multi-encoder structure and a BHTP module, effectively extracting complex SOH features and enhancing stability with a combined pre-training and fine-tuning strategy. The experimental results on the NASA 18650 dataset show that the RMSE and MAE of the B0007 dataset are reduced by 0.020 on average, the RMSE of the B0033 dataset is reduced by 0.005, and the MAE is reduced by 0.003. The method has a significant improvement in RMSE and MAE. The reliability and efficiency of the method are verified.

However, the accuracy of the model for battery SOH prediction will be affected under extreme operating conditions such as high temperature, low temperature, or ultra-fast charging and discharging. In the future, we hope to conduct research in the following key directions. In order to enhance the ability of the model to capture the dynamic characteristics of lithium-ion battery's SOH, the dynamic feature extraction technology based on VAE will be deeply explored, and the optimized network structure will be constructed to accurately capture the dynamics and track the real-time status. In view of the requirements of few-shot scenarios, MAML-based methods will be deeply studied to accelerate model learning and improve prediction accuracy. In this way, the accuracy of SOH prediction in practical scenarios can be further improved. Through the exploration of these research directions, it is expected to further optimize the method so that it can be better applied to practical scenarios.

Author Contributions: Conceptualization, Z.M.; methodology, S.W.; software, S.G. and C.L.; validation, S.G., C.L. and Z.M.; formal analysis, Y.Q.; investigation, Y.Q.; resources, S.W.; data curation, C.L.; writing—original draft preparation, C.L. and S.G.; writing—review and editing, C.L. and Z.M.; visualization, C.L.; supervision, Z.M.; project administration, S.W. All authors have read and agreed to the published version of the manuscript.

Funding: This research was funded by Flying Feather Trail: The Design and Implementation of an Intelligent Cloud System for Wetland Bird Populations, project number 202410128009; Research on the Key Technologies of Dynamically Reconfigurable Battery Modules Based on Digital Twins, project number STZX202307; and Optimization of an Independent Operation DC Microgrid Configuration Scheme and Coordinated Control Study, project number 51867020.

Data Availability Statement: The data that has been used are confidential.

Acknowledgments: All authors thank the editors and reviewers for their attention to the paper. We kindly request the editors and reviewers to provide valuable comments and corrections on this study.

Conflicts of Interest: The authors declare that they have no known competing financial interests or personal relationships.

References

1. Hong, J.; Li, K.; Liang, F.; Yang, H.; Zhang, C.; Yang, Q.; Wang, J. A novel state of health prediction method for battery system in real-world vehicles based on gated recurrent unit neural networks. *Energy* **2024**, *289*, 129918. [CrossRef]
2. Wei, Y.; Wu, D. State of health and remaining useful life prediction of lithium-ion batteries with conditional graph convolutional network. *Expert Syst. Appl.* **2024**, *238*, 122041. [CrossRef]
3. Zhang, X.Y.; Wu, J.H.; Zhang, L.P.; Dai, H.D.; Bai, W.Q.; Lin, H.J.; Zhang, F.; Yang, Y.X. Research progress of lithiumion battery health state assessment method. *Henan Sci.* **2024**, *42*, 1717–1740.
4. Buchanan, S.; Crawford, C. Probabilistic lithium-ion battery state-of-health prediction using convolutional neural networks and Gaussian process regression. *J. Energy Storage* **2024**, *76*, 109799. [CrossRef]
5. Wang, X.; Hu, B.; Su, X.; Xu, L.; Zhu, D. State of health estimation for lithium-ion batteries using random forest and gated recurrent unit. *J. Energy Storage* **2024**, *76*, 109796. [CrossRef]
6. Che, Y.; Zheng, Y.; Forest, F.E.; Sui, X.; Hu, X.; Teodorescu, R. Predictive health assessment for lithium-ion batteries with probabilistic degradation prediction and accelerating aging detection. *Reliab. Eng. Syst. Saf.* **2024**, *241*, 109603. [CrossRef]
7. Wu, T.Z.; Zhu, J.C.; Cheng, X.F.; Kang, J. SOH estimation method for lithium battery Based on charging phase data and GWO-BiLSTM model. *Power Supply Technol.* **2024**, *48*, 2184–2194.
8. Liu, P.; Li, Z.W.; Cai, Y.S.; Wang, W.; Xia, X.Y. Based on the equivalent circuit model and data driven fusion model of SOC and SOH joint estimation approach. *J. Electrotech.* **2024**, *39*, 3232–3243.
9. Wu, L.X.; Pang, H.; Jin, J.M.; Geng, Y.F.; Liu, K. Review of state-of-charge estimation methods for lithium-ion batteries based on electrochemical models. *Trans. China Electrotech. Soc.* **2022**, *37*, 1703–1725.
10. Jin, J.X.; Yu, R.X.; Liu, G.; Xu, L.B.; Ma, Y.Q.; Wang, H.B.; Hu, C. Research progress on estimation methods for the state of health of lithium-ion batteries. *J. Eng. Eng.* **2024**, *19*, 33–48.
11. Messing, M.; Shoa, T.; Habibi, S. Estimating battery state of health using electrochemical impedance spectroscopy and the relaxation effect. *J. Energy Storage* **2021**, *43*, 103210. [CrossRef]
12. Shi, H.S.; Sun, X.W.; Wang, K. Health state estimation of lithium-ion battery based on electrochemical impedance spectroscopy. *Power Gener. Technol.* **2024**, 1–15. Available online: http://kns.cnki.net/kcms/detail/33.1405.tk.20240716.1502.008.html (accessed on 8 March 2025).
13. Luo, F.; Huang, H.; Ni, L.; Li, T. Rapid prediction of the state of health of retired power batteries based on electrochemical impedance spectroscopy. *Chin. J. Sci. Instrum.* **2021**, *42*, 172–180.
14. Wang, Y.; Wei, Q.G.; Sun, P.; Zhang, X.F.; Liu, Y.J. Joint estimation of lithium-battery SOC and SOH based on dual extended kalman filter. In Proceedings of the 2022 China Society of Automotive Engineers Annual Conference, Nantong, China, 20–22 December 2022; pp. 4–9.
15. Yang, D.; Zhang, X.; Pan, R.; Wang, Y.; Chen, Z. A novel gaussian process regression model for state-of-health estimation of lithium-ion battery using charging curve. *J. Power Sources* **2018**, *384*, 387–395. [CrossRef]
16. Hui, Z.L.; Wang, R.J.; Feng, N.N.; Yang, M. State of health prediction of lithium-ion batteries based on ensemble gaussian process regression. *J. Meas. Sci. Instrum.* **2024**, *15*, 397–407. [CrossRef]

17. Tan, Y.; Zhao, G. Transfer learning with long short-term memory network for state-of-health pre-diction of lithium-ion batteries. *IEEE Trans. Ind. Electron.* **2019**, *67*, 8723–8731. [CrossRef]
18. Li, K.; Hu, L.; Song, T.T. State of health estimation of lithium-ion batteries based on CNN-Bi-LSTM. *Shandong Electr. Power Technol.* **2023**, *50*, 66–72.
19. Ke, H. SOH estimation of lithium battery based on diffusion model and bidirectional long short-term memory network. *Ploidy Henan Sci. Technol.* **2024**, *51*, 5–11.
20. Zheng, W.B.; Zhou, X.Y.; Wu, Y.; Feng, L.; Yin, H.T.; Fu, P. Experimental design for state of health (SOH) estimation of lithium batteries based on transfer learning. *Exp. Technol. Manag.* **2022**, *39*. [CrossRef]
21. Yin, J.; Liu, B.; Sun, G.B.; Qian, X.W. Remaining useful life prediction of lithium-ion batteries based on transfer learning and denoising autoencoder-long short-term memory. *Trans. China Electrotech. Soc.* **2024**, *39*, 289–302.
22. Mo, Y.M.; Yu, Z.H.; Ye, P.; Fan, W.J.; Lin, Y. SOH estimation of lithium battery based on transfer learning and GRU neural network. *Acta Sol. Energy* **2024**, *45*, 233–239.
23. Cheng, G.; Wang, X.; He, Y. Remaining useful life and state of health prediction for lithium batteries based on empirical mode decomposition and a long and short memory neural network. *Energy* **2021**, *232*, 121022. [CrossRef]
24. Zhang, X.F.; Chen, Y.L.; Li, S.Q.; Zeng, X.K.; Lian, X.; Huang, C. LSTM neural network based state estimation of lithium ion battery health. *Auto Pract. Technol.* **2025**, *50*, 1–6. [CrossRef]
25. Chang, C.; Wang, Q.; Jiang, J.; Wu, T. Lithium-ion battery state of health estimation using the incremental capacity and wavelet neural networks with genetic algorithm. *J. Energy Storage* **2021**, *38*, 102570. [CrossRef]
26. Huang, J.Y.; Bai, J.Q.; Xiang, Y.H. SOH estimation method for lithium-ion battery based on DOD-LN-GPR model. *Chin. J. Sol. Energy* **2025**, *46*, 60–69.
27. Tian, A.N.; Yang, C.; Gao, Y.; Li, T.Y.; Wang, L.J.; Chang, C.; Jiang, J.C. A state of health estimation method of lithium-ion batteries based on DT-IC-V health features extracted from partial charging segment. *Int. J. Green Energy* **2023**, *20*, 997–1011. [CrossRef]
28. Li, Z.L.; Qiao, G.Z.; Cui, F.S.; Cai, J.H.; Shi, Y.H.; Wang, B.H. Prediction of state of health of lithium-ion batteries using the VMD-HPO-NBEATS model. *China Test* **2024**, *50*, 65–73.
29. Shi, S.S.; Gao, Z.B. Based on the automatic feature extraction and IWOA—Lithium-ion battery SOH prediction of SVR model. *Intern. Combust. Engines Accessories* **2024**, *22*, 10–13.
30. Li, J.; Chen, X.L.; Xu, L. Li-ion battery health estimation based on multiple health factors and IPSO-LSTM model. *Automot. Engine* **2025**, *1*, 39–46.
31. Zhang, M.; Yang, D.F.; Du, J.X.; Sun, H.L.; Li, L.W.; Wang, L.C.; Wang, K. A Review of SOH prediction of li-ion batteries based on data-driven algorithms. *Energies* **2023**, *16*, 3167. [CrossRef]
32. Dong, X.H.; Dong, J.B.; Wang, M.S.; Zeng, F.; Pan, Y. Rapid estimation method of lithium battery health state Based on new health characteristics. *Electr. Power Eng. Technol.* **2025**, *44*, 136–142+206.
33. Shi, Y.H.; Tian, J.Y.; Liu, J.J.; Yang, S.Q. Research on multi-source adaptive transfer learning algorithm based on manifold structure and its application. *Control. Decis.* **2023**, *38*, 797–804.
34. Li, L.; Yan, X.M.; Zhang, Y.S.; Feng, Y.L.; Hu, H.L.; Duan, Y.J.; Cui, C.H. Lithium battery state of charge estimation method based on transfer learning. *J. Xi'an Jiao Tong University* **2023**, *57*, 142–150.
35. Liu, B.; Yang, J.; Wang, R.G.; Xue, L.X. Memory-based transfer learning for few-shot learning. *Comput. Eng. Appl.* **2022**, *58*, 242–249.
36. Chang, W.; Hu, Z.C.; Pan, D.Z.; Shi, J.W. Application of an improved VAE-GAN model to battery EIS data augmentation. *Sci. Technol. Ind.* **2024**, *24*, 258–263.
37. Zhao, Y.P.; Huang, W.; Zhang, J.F. Lithium battery life prediction based on LSTM-Transformer with hybrid scale health factor. *Electron. Meas. Technol.* **2024**, *11*, 112–122.
38. Li, D.C.; Wu, C.S.; Tsai, T.I.; Lina, Y.S. Using mega-trend-diffusion and artificial samples in small data set learning for early flexible manufacturing system scheduling knowledge. *Comput. Oper. Res.* **2007**, *34*, 966–982. [CrossRef]
39. Qiao, J.F.; Guo, Z.H.; Tang, J. Virtual sample generation method based on improved metertrend diffusion and hidden layer interpolation and its application. *Chin. Chem. Eng. J.* **2020**, *71*, 5681–5695.
40. Kang, G.; Wu, L.; Guan, Y.; Peng, Z. A Virtual sample generation method based on differential evolution Al-gorithm for overall trend of small sample data: Used for lithium-ion battery capacity degradation data. *IEEE Access* **2019**, *7*, 123255–123267. [CrossRef]
41. Zhu, H.J.; Lü, Z.G.; Di, R.H.; Sun, X.J.; Hao, K.Q. To improve MD-MTD lithium battery life prediction neural network simulation. *J. Xi'an Univ. Technol.* **2022**, *42*, 620–626.
42. Xu, Z.J.; Zhu, Q.; Xu, S.F.; Mao, Q. Research on parameter identification algorithm of lithium-battery model based on differential evolution method. *Electr. Eng.* **2021**, *7*, 35–37.
43. Zhu, Y.F.; Jiang, G.P.; Gao, H.; Li, W.Z.; Gui, Y.C. Prediction of battery state of charge based on feature selection and data augmentation. *Comput. Syst. Appl.* **2023**, *32*, 45–54.

44. Cui, X.K.; Wang, Q.Z.; Liu, Q.P. State of health prediction of lithium batteries based on data-driven approaches. *Complex Syst. Complex. Sci.* **2024**, *21*, 154–159.
45. Liu, X.H.; Gao, Z.C.; Tian, J.Q.; Wei, Z.B.; Fang, C.Q.; Wang, P. State of health estimation for lithium-ion batteries using voltage curves reconstruction by conditional generative adversarial network. *IEEE Trans. Transp. Electrif.* **2024**, *10*, 10557–10567. [CrossRef]
46. Ren, Y.; Tang, T.; Jiang, F.S.; Xia, Q.; Zhu, X.Y.; Sun, B.; Yang, D.Z.; Feng, Q.; Qian, C. A novel state of health estimation method for lithium-ion battery pack based on cross generative adversarial networks. *Appl. Energy* **2025**, *377*, 124385. [CrossRef]
47. Guo, F.; Huang, G.S.; Zhang, W.C.; Wen, A.; Li, T.T.; He, H.C.; Huang, H.L.; Zhu, S.S. Lithium battery state-of-health estimation based on sample data generation and temporal convolutional neural network. *Energies* **2023**, *16*, 8010. [CrossRef]
48. Lin, C.H.; Kaushik, C.; Dyer, E.L.; Muthukumar, V. The good, the bad and the ugly sides of data augmentation: An implicit spectral regularization perspective. *J. Mach. Learn. Res.* **2024**, *25*, 1–85.
49. Wang, Q.H.; Jia, H.J.; Huang, L.X.; Mao, Q.R. Semantic contrastive clustering with federated data augmentation. *J. Comput. Res. Dev.* **2024**, *61*, 1511–1524.
50. Thompson, M.L.; Kramer, M.A. Modeling chemical processes using prior knowledge and neural networks. *Alche J.* **1994**, *40*, 1328–1340. [CrossRef]
51. Zhao, W.; Yan, H.; Zhou, Z. Transformer fault diagnosis based on a residual BP neural network. *Electr. Power Autom. Equip.* **2020**, *310*, 143–148.
52. Hu, X.Q.; Geng, L.M.; Shu, J.H.; Zhang, W.B.; Wu, C.L.; Wei, X.L.; Huang, D.; Chen, H. Health state estimation of lithium-ion battery using global health factor and residual Model. *J. Xi'an Jiaotong Univ.* **2025**, *59*, 105–117.
53. Hao, K.Q.; Lü, Z.G.; Li, Y.; Di, R.H.; Zhu, H.J. Remaining useful life prediction of lithium-ion batteries using BP neural network incorporating prior knowledge. *J. Xi'an Technol. Univ.* **2022**, *42*, 65–73.
54. Fan, J.H.; Liu, Q.Y.; Ma, L.; Liu, L.H. Life prediction model of lithium-ion batteries for electric vehicles based on deep learning. *J. Terahertz Sci. Electron. Inf. Technol.* **2025**, *23*, 182–187.
55. Jiang, B.; Zhu, J.G.; Wang, X.Y.; Wei, X.Z.; Shang, W.L.; Dai, H.F. A comparative study of different features extracted from electrochemical impedance spectroscopy in state of health estimation for lithium-ion batteries. *Appl. Energy* **2022**, *322*, 119502. [CrossRef]
56. Li, J.H.; Li, T.T.; Qiao, Y.J.; Tan, Z.J.; Qiu, X.H.; Deng, H.; Li, W.; Qi, X.; Wu, W.X. Internal temperature estimation method for lithium-ion battery based on multi-frequency imaginary part impedance and GPR model. *J. Energy Storage* **2025**, *118*, 116287. [CrossRef]
57. Li, Z.H.; Shi, Q.L.; Wang, K.L.; Jiang, K. Research status and prospect of lithium-ion battery health state estimation methods. *Autom. Electr. Power Syst.* **2024**, *48*, 109–129.
58. Zhou, Y.F.; Sun, X.X.; Huang, L.J.; Lian, J. Life cycle oriented health state estimation of lithium battery. *J. Harbin Inst. Technol.* **2021**, *53*, 55–62.
59. Zhang, S.F.; Zhang, Q.Y.; Yang, Y.S.; Su, Y.X.; Xiong, B.Y. Modeling method of lithium-ion battery based on sliding window and LSTM neural network. *Energy Storage Sci. Technol.* **2022**, *11*, 228–239.
60. Chen, M.H.; Wang, T.; Yuan, Y.; Ke, S.T. Study on retinal OCT segmentation with dual-encoder. *Opto-Electron. Eng.* **2023**, *50*, 230146.
61. Ju, F.J.; Wu, Y.C.; Dong, M.J.; Zhao, J.X. MiFDeU: Multi-information fusion network based on dual-encoder for pelvic bones segmentation. *Eng. Appl. Artif. Intell.* **2025**, *147*, 110230. [CrossRef]
62. NASA Ames Prognostics Data Repository. Available online: https://www.nasa.gov/intelligent-systems-division/discovery-and-systems-health/pcoe/pcoe-data-set-repository/ (accessed on 10 April 2025).
63. Van Houdt, G.; Mosquera, C.; Nápoles, G. A review on the long short-term memory model. *Artif. Intell. Rev.* **2020**, *53*, 5929–5955. [CrossRef]
64. Huang, Z.; Xu, W.; Yu, K. Bidirectional LSTM-CRF models for sequence tagging. *arXiv* **2015**, arXiv:1508.01991.
65. Zargar, S. Introduction to Sequence Learning Models: RNN, LSTM, GRU. 2021. Available online: https://www.researchgate.net/profile/Sakib-Zargar-2/publication/350950396_Introduction_to_Sequence_Learning_Models_RNN_LSTM_GRU/links/607b41c0907dcf667ba83ade/Introduction-to-Sequence-Learning-Models-RNN-LSTM-GRU.pdf (accessed on 8 March 2025).

MDPI AG
Grosspeteranlage 5
4052 Basel
Switzerland
Tel.: +41 61 683 77 34

Batteries Editorial Office
E-mail: batteries@mdpi.com
www.mdpi.com/journal/batteries

www.ingramcontent.com/pod-product-compliance
Lightning Source LLC
Chambersburg PA
CBHW062037220926
51850CB00010B/2586
* 9 7 8 3 7 2 5 8 6 1 9 9 6 *